T0325845

Five-Dimensional Physics

Classical and Quantum Consequences
of Kaluza-Klein Cosmology

Paul S Wesson

University of Waterloo, Canada & Stanford University, USA

Five-Dimensional Physics

Classical and Quantum Consequences of Kaluza-Klein Cosmology

 **World Scientific**

NEW JERSEY · LONDON · SINGAPORE · BEIJING · SHANGHAI · HONG KONG · TAIPEI · CHENNAI

Published by

World Scientific Publishing Co. Pte. Ltd.

5 Toh Tuck Link, Singapore 596224

USA office: 27 Warren Street, Suite 401-402, Hackensack, NJ 07601

UK office: 57 Shelton Street, Covent Garden, London WC2H 9HE

British Library Cataloguing-in-Publication Data
A catalogue record for this book is available from the British Library.

FIVE-DIMENSIONAL PHYSICS
Classical and Quantum Consequences of Kaluza-Klein Cosmology

ISBN-13 978-981-256-661-4
ISBN-10 981-256-661-9

Printed in Singapore

PREFACE

Five dimensions represents a unique situation in modern theoretical physics. It is the simplest extension of the four-dimensional Einstein theory of general relativity, which is the basis of astrophysics and cosmology. It is also widely regarded as the low-energy limit of higher-dimensional theories which seek to unify gravity with the interactions of particle physics. In the latter regard, we can mention 10D supersymmetry, 11D supergravity and higher-D string theory. However, the view of our group is pragmatic: we need to understand 5D physics, to put 4D gravity into perspective and to show us where to go in higher dimensions.

This book provides an account of the main developments in 5D physics in recent years. In a sense, it is a sequel to the omniverous volume *Space-Time-Matter* published in 1999. However, the present account is self-contained. So are the chapters, which each deal with a separate topic and has its own bibliography. The major topics are cosmology, quantum physics and embeddings. There are currently two approaches to these topics, namely those provided by induced-matter theory and membrane theory. The former uses the fifth dimension in an unrestricted manner, to provide an explanation for the mass-energy content of the universe. The latter uses the fifth dimension to define a hypersurface, to which the interactions of particle physics are confined while gravity propagates freely into the "bulk". Physically, these two versions of 5D physics are differently motivated, but mathematically they are equivalent (one can always insert a membrane into the former to obtain the latter). Therefore, in order to be general, this volume concentrates on the mathematical formalism.

Some knowledge of tensor calculus is presumed, but each chapter starts and ends with a qualitative account of its contents. Many of the results presented here are the result of a group effort. Thanks are due to the senior researchers whose work is described herein, notably H. Liu, B. Mashhoon and J. Ponce de Leon. Acknowledgements should also be made to associates from various fields including T. Fukui, P. Halpern and J.M. Overduin. Gratitude is further owed to a cadre of enthusiastic graduate students, notably D. Bruni, T. Liko and S.S. Seahra. Much of this book was written during a stay with Gravity-Probe B of the Hansen Physics Laboratories at Stanford University, at the invitation of C.W.F. Everitt. Any omissions or errors are the responsibility of the author.

Theoretical physics can be an arcane and even boring subject. However, the author is of the opinion that the fifth dimension is fascinating. Where else can one discover that the universe may be flat as viewed from higher dimensions, or that spacetime uncertainty is the consequence of deterministic laws in a wider world? Such issues provide a healthy shake to the bedrock of conventional physics, dislodging the plastered-over parts of its edifice and providing a stronger foundation for future work. Physics and philosophy are not, it appears, separate. This book provides technical results whose success leads inevitably to the insight that there is more to the world than is apparent, provided one looks...

Paul S. Wesson

CONTENTS

1. HIGHER-DIMENSIONAL PHYSICS

"There's more to this than meets the eye" (Old English saying)

1.1   Introduction

Theoretical physics is in the happy situation of being able to pluck good ideas from philosophy, work them through using the machinery of algebra, and produce something which is both stimulating and precise. It goes beyond words and equations, because when properly done it encapsulates what many people regard as reality.

We sometimes tend to forget what a stride was made when Newton realized that the force which causes an apple to fall to the ground is the same one which keeps the Moon in its orbit – and which is now known to influence the motions of even the most remote galaxies. Nowadays, gravity has to be considered in conjunction with electromagnetism plus the weak and strong forces of particle physics. Even so, it is still possible to give an account of modern physics in a few hundred pages or so. On reflection, this is remarkable. It comes about because of the enormous efficiency of mathematics as the natural language of physics, coupled with the tradition whereby physicists introduce the least number of hypotheses necessary to explain the natural world (Occam's razor of old). At present, it is commonly believed that the best way to explain all of the forces of physics is via the idea of higher dimensions.

In this regard, five-dimensional field theory is particularly useful, as it is the basic extension of the four-dimensional spacetime

1

of Einstein gravity and is widely regarded as the low-energy limit of higher-dimensional theories which more fully address the particle interactions. This slim volume is a concise account of recent developments in 5D theory and their implications for classical and quantum physics.

1.2   Dimensions Then and Now

The idea of a "dimension" is primitive and at least partly intuitive. Recent histories of the idea are given in the books by Wesson (1999) and Halpern (2004). It was already established by the time of Newton, who realized that mass was a more fundamental concept than density, and that a proportionality between physical quantities could be converted to an equation if the latter balanced its ingredients of mass, length and time (i.e., was dimensionally homogeneous). Hence the introduction of a parameter G, which we now call Newton's constant of gravity.

The coordinates of an object (x, y, z) in ordinary space and that of local time (t) are, of course, the basic dimensions of geometry. But the concept of force, at least the gravitational kind, obliges us to introduce another dimension related to the mass of an object (m). And modern physics recognizes other such, notably the one which measures a body's electric charge (q). The role of the so-called fundamental constants of physics is primarily to transpose quantities like mass and charge into geometrical ones, principally lengths (Wesson 1999, pp. 2-11). This is illustrated most cogently by the conversion of the time to an extra coordinate $x^4 \equiv ct$ via the use of the speed of light, a ploy due to Minkowski and Einstein which forms the foundation of 4D spacetime.

The idea of a dimension is, to a certain extent, malleable. It is also important to notice that modern field theories, like general relativity, are written in terms of tensor equations which are not restricted in their dimensionality. One can speculate that had Einstein been formulating his theory of gravity today, he might have established this anonymity of dimension as a principle, on a par with the others with which we are familiar, such as that of equivalence (see Chapter 3). It is this freedom to choose the dimensionality which underlies the numerous extensions of general relativity. These include the original 5D Kaluza-Klein theory, its modern variants which are called induced-matter and membrane theory, plus the higher extensions such as 10D supersymmetry, 11D supergravity and the higher-D versions of string theory.

Kaluza initiated field theory with more than the 4 dimensions of spacetime in 1921, when he published a paper which showed how to unify gravity (as described by Einstein's equations) with electromagnetism (as described by Maxwell's equations). It is well known that Einstein kept Kaluza's paper for a couple of years before finally as referee allowing it to go forward. However, Einstein was then and remained in his later years an advocate of extra dimensions. For example, a letter to Kaluza from Einstein in 1919 stated "The formal unity of your theory is astonishing" (Halpern, 2004, p.1). Indeed, the natural way in which the 4 Maxwell equations fall out of the 15 field equations of what is a kind of general relativity in 5D, has since come to be called the Kaluza-Klein miracle. However, the mathematical basis of the unification is simple: In 5D there are 15 independent components of the metric tensor, of which one refers to a scalar field which was not at the time considered significant and was so suppressed. For similar reasons, to do with the presumed unobservability of effects to do with the extra dimension, all derivatives of the other metric coefficients with respect to the extra coordinate were set to zero (the "cylinder" condition). This left 14 metric coefficients, which could depend on the 4 coordinates of spacetime (x^α, $\alpha = 0,123$ for t, xyz). These 14 coefficients were determined by 14 field equations. The latter turned out to be the 10 Einstein equations and the 4 Maxwell equations. Voila: a unification of gravity and electromagnetism.

Klein pushed the 5D approach further in 1926, when he published a paper which showed how to incorporate quantum effects into the theory. He did this by the simple device of assuming that the topology of the extra dimension was not flat and open, but curved into a circle. In other words, while a local orbit in spacetime (x^α) would be straight, an orbit in the extra dimension (x^4) would merely go around and around. This cyclic behaviour would lead to quantum effects, *provided* the extra dimension were rolled up to a microscopic size ("compactification"). The size of the extra dimension was presumed to be related to the parameter typical of quantum phenomena, namely

Planck's constant h. Among other consequences of the closed topology of the extra dimension, it was shown that the cyclic momentum could be related to the charge of the electron e, thus explaining its quantization.

The brainwaves of Kaluza and Klein just summarized are the kind which are neat and yet powerful. They continued to be held in high regard for many years in theoretical physics, even though the latter was redirected by the algebraically simple and effective ideas on wave mechanics that were soon introduced by Schrodinger, Heisenberg and Dirac. Kaluza-Klein theory later underwent a revival, when Einstein's theory was recognized as the best basis for cosmology. But something has to be admitted: Kaluza-Klein theory in its original form is almost certainly wrong.

By this, it is not meant that an experiment was performed which in the standard but simplistic view of physics led to a disproval of the 5D theory. Rather, it means that the original Kaluza-Klein theory is now acknowledged as being at odds with a large body of modern physical lore. For example, the compactification due to Klein leads to the prediction that the world should be dominated by particles with the Planck mass of order 10^{-5} g, which is clearly not the case. (This mismatch is currently referred to as the hierarchy problem, to which we will return.) Also, the suppression of the scalar field due to Kaluza leaves little room to explain the "dark energy" currently believed to be a major component of the universe. (This is a generic form of what is commonly referred to as the cosmological-constant problem, to which attention will be given later.) Further, the cylinder condition assumed by both fathers of 5D field theory effectively rules out any way to explain matter as a geometrical effect, something which Einstein espoused and is still the goal of many physicists.

It is instructive to recall at this juncture the adage which warns us not to throw out the baby with the bath-water. In this instance, the baby is the concept of a 5 (or higher) D space; whereas the water is the smothering algebraic restrictions which were applied to the theory in its early days as a means of making progress, but which are now no longer needed. Hence modern Kaluza-Klein theory, which is algebraically rich and exists in several versions.

1.3   Higher-Dimensional Theories

These may be listed in terms of their dimensionality and physical motivation. However, all are based on Einstein's theory of general relativity. The equations for this and its canonical extension will be deferred to the next section.

Induced-matter theory is based on an unrestricted 5D manifold, where the extra dimension and derivatives with respect to the extra coordinate are used to explain the origin of 4D matter in terms of geometry. (For this reason, it is sometimes called space-time-matter theory.) As mentioned above, this goal was espoused by Einstein, who wished to transpose the "base-wood" of the right-hand side of his field equations into the "marble" of the left-hand side. That is, he wished to find an algebraic expression for what is usually called the energy-momentum tensor $(T_{\alpha\beta})$, which was on the same footing as the purely geometrical object we nowadays refer to as the Einstein tensor $(G_{\alpha\beta})$. That this is possible in practice was proved using an algebraic reduction of the 5D field equations by Wesson and Ponce de Leon (1992). They were, however, unaware that the technique was guaranteed in principle by a little-known theorem on local embeddings of Riemannian manifolds by Campbell (1926). We will return to the field equations and their embeddings below. Here, we note that the field equations of 5D relativity with a scalar field and dependence on the extra coordinate in general lead to 15 second-order, non-linear relations. When the field equations are set to zero to correspond to a 5D space which is apparently empty, a subset of them gives back the 10 Einstein field equations in 4D *with sources*. That is, there is an effective or induced 4D energy-momentum tensor which has the properties of what we normally call matter, but depends on the extra metric coefficients and derivatives with respect to the extra coordinate. The other 5 field equations give back a set of 4 Maxwell-like or conservation equations, plus 1 scalar relation which has the form of a wave equation. Following the demonstration that matter could be viewed as a consequence of geometry, there was a flurry of activity, resulting in several theorems and numerous exact solutions (see Wesson 1999 for a catalog). The theory has a 1-body solution which satisfies all of the classical tests of relativity in astrophysics, as well as other solutions which are relevant to particle physics.

Membrane theory is based on a 5D manifold in which there is a singular hypersurface which we call 4D spacetime. It is motivated by the wish to explain the apparently weak strength of gravity as compared to the forces of particle physics. It does this by assuming that gravity propagates freely (into the 5D bulk), whereas particle interactions are constrained to the hypersurface (the 4D brane). That this is a practical approach to unification was realized by Randall and Sundrum (1998, 1999) and by Arkani-Hamed, Dimopoulos and Dvali (1998, 1999). The original theory helped to explain the apparently small masses of elementary particles, which is also referred to as the hierarchy problem. In addition, it helped to account for the existence and size of the cosmological constant, since that parameter mediates the exponential factor in the extra coordinate which is typical of distances measured away from the brane. As with induced-matter theory, the membrane approach has evolved somewhat since its inception. Thus there has been discussion of thick branes, the existence of singular or thin branes in (4+d) dimensions or d-branes, and the possible collisions of branes as a means of explaining the big bang of traditional 4D cosmology. It should also be mentioned that the field equations of induced-matter and membrane theory have recently been shown to be equivalent by Ponce de Leon (2001; see below also). This means that the implications of these approaches for physics owes more to interpretation than algebra, and exact solutions for the former theory can be carried over to the latter.

Theories in $N > 5$ dimensions have been around for a considerable time and owe their existence to specific physical circumstances. Thus 10D supersymmetry arose from the wish to pair every integral-spin boson with a half-integral-spin fermion, and thereby cancel the enormous vacuum or zero-point fields which would ensue otherwise. The connection to ND classical field theory involves the fact that it is possible to embed any *curved* solution (with energy) of a 4D theory in a *flat* solution (without energy) of a higher-dimensional theory, provided the larger manifold has a dimension of $N \geq 10$. From the viewpoint of general relativity or a theory like it, which has 10 independent components of the metric tensor or potentials, this is hardly surprising. The main puzzle is that while supersymmetry is a property much to be desired from the perspective of theoretical particle physics, it must be very badly broken in a practical sense. The reason for this apparent conflict between theory and practice may have to do with our (perhaps unjustified) wish to reduce physics to 4D, and/or our (probably incomplete) knowledge of how to categorize the properties of particles using internal symmetry groups. The latter have, of course, to be taken into account when we attempt to estimate the "size" of the space necessary to accommodate both gravity and the particle interactions. Hence the possible unification in terms of (4+7)D or 11D supergravity. However, a different approach is to abandon completely the notion of a point – with its implied singularity – and instead model particles as strings (Szabo 2004, Gubser and Lykken 2004). The logic of this sounds compelling, and string theory offers a broad field for development. But line-like singularities are not unknown, and some of the models proposed have an unmanageably high dimensionality (e.g., $N = 26$). One lesson which can be drawn, though, from $N > 5D$ theory is that there is no holy value of N which is to be searched for as if it were a shangri-la of physics. Rather, the value of N is to be chosen on utilitarian grounds, in accordance with the physics to be studied.

1.4   Field Equations in $N \geq 4$ Dimensions

Just as Maxwell's equations provided the groundwork for Einstein's equations, so should general relativity be the foundation for field equations that use more than the 4 dimensions of spacetime.

Einstein's field equations are frequently presented as a match between a geometrical object $G_{\alpha\beta}$ and a physical object $T_{\alpha\beta}$, via a coupling constant κ, in the form $G_{\alpha\beta} = \kappa T_{\alpha\beta} (\alpha, \beta = 0,1,2,3)$. Here the Einstein tensor $G_{\alpha\beta} \equiv R_{\alpha\beta} - R g_{\alpha\beta}/2$ depends on the Ricci tensor, the Ricci scalar and the metric tensor, the last defining small intervals in 4D by a quadratic line element $ds^2 = g_{\alpha\beta} dx^\alpha dx^\beta$. The energy-momentum tensor $T_{\alpha\beta}$ depends conversely on common properties of matter such as the density ρ and pressure p, together with the 4-velocities $u^\alpha \equiv dx^\alpha / ds$. However, even Einstein realized that this split between geometry and matter is subjective and artificial. One example of this concerns the cosmological constant Λ. This was originally added to the left-hand side of the field equations as a geo-metrical term $\Lambda g_{\alpha\beta}$, whence the curvature it causes in spacetime cor-responds to a force per unit mass (or acceleration) $\Lambda r c^2 / 3$, where r is the distance from a suitably chosen origin of coordinates. But nowadays, it is commonly included in the right-hand side of the field equations as an effective source for the vacuum, whose equation of state is $p_v = -\rho_v c^2$, where $\rho_v = \Lambda c^2 / 8\pi G$ corresponds to the den-sity of a non-material medium. (Here we take the dimensions of Λ as length^{-2} and retain physical units for the speed of light c and gravi-tational constant G, so the coupling constant in the field equations is $\kappa = 8\pi G / c^4$.) The question about where to put Λ is largely one of semantics. It makes little difference to the real issue, which is to ob-tain the $g_{\alpha\beta}$ or potentials from the field equations.

The latter in traditional form are

$$R_{\alpha\beta} - \frac{1}{2} R g_{\alpha\beta} + \Lambda g_{\alpha\beta} = \frac{8\pi G}{c^4} T_{\alpha\beta} \qquad . \qquad (1.1)$$

Taking the trace of this gives $R = 4\Lambda - (8\pi G / c^4) T$ where $T \equiv g^{\alpha\beta} T_{\alpha\beta}$. Using this to eliminate R in (1.1) makes the latter read

$$R_{\alpha\beta} = \frac{8\pi G}{c^4} \left(T_{\alpha\beta} - \frac{1}{2} T g_{\alpha\beta} \right) + \Lambda g_{\alpha\beta} \qquad . \qquad (1.2)$$

Here the cosmological constant is treated as a source term for the vacuum, along with the energy-momentum tensor of "ordinary" matter. If there is none of the latter then the field equations are

$$R_{\alpha\beta} = \Lambda g_{\alpha\beta} \qquad . \qquad (1.3)$$

These have 10 independent components (since $g_{\alpha\beta}$ is symmetrical). They make it clear that Λ measures the mean radius of curvature of a 4D manifold that is empty of conventional sources, i.e. vacuum. If there are no sources of *any* kind – or if the ordinary matter and vacuum fields cancel as required by certain symmetries – then the field equations just read

$$R_{\alpha\beta} = 0 \qquad . \qquad (1.4)$$

It is these equations which give rise to the Schwarzschild and other solutions of general relativity and are verified by observations.

The field equations of 5D theory are taken by analogy with (1.4) to be given by

$$R_{AB} = 0 \ \left(A, B = 0,123,4\right) \qquad . \qquad (1.5)$$

Here the underlying space has coordinates $x^A = (t, xyz, l)$ where the last is a length which is commonly taken to be orthogonal to spacetime. The associated line element is $dS^2 = g_{AB} dx^A dx^B$, where the 5D metric tensor now has 15 independent components, as does (1.5). However, the theory is covariant in its five coordinates, which may be chosen for convenience. Thus a choice of coordinate frame, or gauge, may be made which reduces the number of g_{AB} to be determined from 15 to 10. This simplifies both the line element and the field equations.

The electromagnetic gauge was used extensively in earlier work on 5D relativity, since it effectively separates gravity and electromagnetism. A more modern form of this expresses the 5D line element as parts which depend on $g_{\alpha\beta}$ (akin to the Einstein gravitational potentials with associated interval $ds^2 = g_{\alpha\beta} dx^\alpha dx^\beta$), Φ (a scalar field which may be related to the Higgs field by which particles acquire their masses), and A_μ (related to the Maxwell potentials of classical electromagnetism). The 5D line element then has the form

$$dS^2 = ds^2 + \varepsilon\,\Phi^2 \left(dx^4 + A_\mu dx^\mu \right)^2 \quad . \tag{1.6}$$

Here $\varepsilon = \pm 1$ determines whether the extra dimension $\left(g_{44} = \varepsilon\,\Phi^2 \right)$ is spacelike or timelike: both are allowed by the mathematics, and we will see elsewhere that $\varepsilon = -1$ is associated with particle-like behaviour while $\varepsilon = +1$ is associated with wave-like behaviour. Most work has been done with the former choice, so we will often assume that the 5D metric has signature $(+ - - - -)$. Henceforth, we will also absorb the constants c and G by a suitable choice of units. Then the dynamics which follows from (1.6) may be investigated by minimizing the 5D interval, via $\delta\left[\int dS\right] = 0$ (Wesson 1999, pp. 129-153). In general, the motion consists of the usual geodesic one found in Einstein theory, plus a Lorentz-force term of the kind found in Maxwell theory, and other effects due to the extended nature of the geometry including the scalar field. Further results on the dynamics, and the effective 4D energy-momentum tensor associated with the off-diagonal terms in line elements like (1.6), have been worked out by Ponce de Leon (2002). We eschew further discussion of metrics of this form, however, to concentrate on a more illuminating case.

The gauge for neutral matter has a line element which can be written

$$dS^2 = g_{\alpha\beta}\left(x^\gamma, l\right) dx^\alpha dx^\beta + \varepsilon\,\Phi^2\left(x^\gamma, l\right) \quad . \qquad (1.7)$$

In this we have set the electromagnetic potentials ($g_{4\alpha}$) to zero, but the remaining degree of coordinate freedom has been held in reserve. (It could in principle be used to flatten the scalar potential via $\left|g_{44}\right| = 1$, but while we will do this below it is instructive to see what effects follow from this field.) The components of the 5D Ricci tensor for metric (1.7) have wide applicability. They are:

$$
{}^5 R_{\alpha\beta} = {}^4 R_{\alpha\beta} - \frac{\Phi_{,\alpha;\beta}}{\Phi} + \frac{\varepsilon}{2\Phi^2}\left(\frac{\Phi_{,4}\,g_{\alpha\beta,4}}{\Phi} - g_{\alpha\beta,44}\right.
$$
$$
\left. + g^{\lambda\mu} g_{\alpha\lambda,4}\,g_{\beta\mu,4} - \frac{g^{\mu\nu} g_{\mu\nu,4}\,g_{\alpha\beta,4}}{2}\right)
$$

$$R_{4\alpha} = \Gamma\left(\frac{g^{\beta\lambda} g_{\lambda\alpha,4} - \delta_\alpha^\beta g^{\mu\nu} g_{\mu\nu,4}}{2\Gamma}\right)_{,\beta} + \frac{g^{\mu\beta} g_{\mu\beta,\lambda} g^{\lambda\sigma} g_{\sigma\alpha,4}}{4}$$

$$-\frac{g^{\lambda\beta} g_{\beta\mu,\alpha} g^{\mu\sigma} g_{\sigma\lambda,4}}{4}$$

$$R_{44} = -\varepsilon\,\Phi\Box\Phi - \frac{g^{\lambda\beta}_{\quad,4} g_{\lambda\beta,4}}{2} - \frac{g^{\lambda\beta} g_{\lambda\beta,44}}{2} +$$

$$\frac{\Phi_{,4} g^{\lambda\beta} g_{\lambda\beta,4}}{2\Phi} - \frac{g^{\mu\beta} g^{\lambda\sigma} g_{\lambda\beta,4} g_{\mu\sigma,4}}{4} \ . \tag{1.8}$$

Here a comma denotes the ordinary partial derivative, a semicolon denotes the ordinary 4D covariant derivative, $\Box\Phi \equiv g^{\mu\nu}\Phi_{,\mu;\nu}$ and $\Gamma \equiv \left|\varepsilon\Phi^2\right|^{1/2}$. Superscripts are used here and below for the 5D tensors and their purely 4D parts, whenever there is a risk of confusion. When the components (1.8) are used with the 5D field equations (1.5), it is clear that we obtain tensor, vector and scalar equations which have distinct applications in physics.

The tensor components of (1.8), in conjunction with the 5D field equations $R_{AB} = 0$ (1.5), give the 10 field equations of Einstein's general relativity. The method by which this occurs is by now well known (Wesson and Ponce de Leon 1992). In summary, we form the conventional 4D Ricci tensor, and with it and the 4D Ricci scalar construct the 4D Einstein tensor $G_{\alpha\beta} \equiv {}^4R_{\alpha\beta} - {}^4R g_{\alpha\beta}/2$. The remaining terms in $^5R_{\alpha\beta}$ of (1.8) are then used to construct an effective or induced 4D energy-momentum tensor via $G_{\alpha\beta} = 8\pi T_{\alpha\beta}$. Several instructive results emerge during this process. For example, the 4D scalar curvature just mentioned may be shown using all of (1.8) to ge given by

$$^4R = \frac{\varepsilon}{4\Phi^2}\left[g^{\mu\nu}{}_{,4}g_{\mu\nu,4} + \left(g^{\mu\nu}g_{\mu\nu,4}\right)^2\right] \quad . \tag{1.9}$$

This relation has been used implicitly in the literature, but explicitly as here it shows that: (a) What we call the curvature of 4D spacetime can be regarded as the result of embedding it in an x^4-dependent 5D manifold; (b) the sign of the 4D curvature depends on the signature of the 5D metric; (c) the magnitude of the 4D curvature depends strongly on the scalar field or the size of the extra dimension $\left(g_{44} = \varepsilon\Phi^2\right)$, so while it may be justifiable to neglect this in astrophysics (where the 4D curvature is small) it can be crucial in cosmology and particle physics. Another instructive result concerns the form of the 4D energy-momentum tensor. It is given by

$$8\pi T_{\alpha\beta} = \frac{\Phi_{,\alpha;\beta}}{\Phi} - \frac{\varepsilon}{2\Phi^2}\left\{\frac{\Phi_{,4}g_{\alpha\beta,4}}{\Phi} - g_{\alpha\beta,44} + g^{\lambda\mu}g_{\alpha\lambda,4}g_{\beta\mu,4}\right.$$

$$\left. - \frac{g^{\mu\nu}g_{\mu\nu,4}g_{\alpha\beta,4}}{2} + \frac{g_{\alpha\beta}}{4}\left[g^{\mu\nu}{}_{,4}g_{\mu\nu,4} + \left(g^{\mu\nu}g_{\mu\nu,4}\right)^2\right]\right\} \quad . \tag{1.10}$$

This relation has been used extensively in the literature, where it has been shown to give back all of the properties of ordinary matter (such as the density and pressure) for standard solutions. However, it has further implications, and shows that: (a) What we call matter in a curved 4D spacetime can be regarded as the result of the embedding in an x^4-dependent (possibly flat) 5D manifold; (b) the nature of the 4D matter depends on the signature of the 5D metric; (c) the 4D source depends on the extrinsic curvature of the embedded 4D spacetime and the scalar field associated with the extra dimension, which while they are in general mixed correspond loosely to ordinary matter and the stress-energy of the vacuum. In conclusion for this paragraph, we see that a 5D manifold – which is apparently empty – contains a 4D manifold with sources, where the tensor set of the 5D field equations corresponds to the 4D Einstein equations of general relativity.

The vector components of (1.8), in conjunction with (1.5), can be couched as a set of conservation equations which resemble those found in Maxwellian electromagnetism and other field theories. They read

$$P^\beta_{\alpha;\beta} = 0 \quad , \tag{1.11}$$

where the 4-vector concerned is defined via

$$P^\beta_\alpha \equiv \frac{1}{2\Phi}\left(g^{\beta\sigma}g_{\sigma\alpha,4} - \delta^\beta_\alpha g^{\mu\nu}g_{\mu\nu,4}\right) \quad . \tag{1.12}$$

These are usually easy to satisfy in the continuous fluid of induced-matter theory, and are related to the stress in the surface ($x^4 = 0$) of membrane theory with the Z_2 symmetry (see below). It should be noted that these relations do not come from some external criterion such as the minimization of the line element, but are derived from and are an inherent part of the field equations.

The scalar or last component of (1.8), when set to zero in accordance with the field equations (1.5), yields a wave-type equation for the potential associated with the fifth dimension $\left(g_{44} = \varepsilon \Phi^2\right)$ in the metric (1.7). It is

$$\Box\Phi = -\frac{\varepsilon}{2\Phi}\left[\frac{g^{\lambda\beta}{}_{,4}g_{\lambda\beta,4}}{2} + g^{\lambda\beta}g_{\lambda\beta,44} - \frac{\Phi_{,4}g^{\lambda\beta}g_{\lambda\beta,4}}{\Phi}\right] \quad (1.13)$$

Here as before $\Box\Phi \equiv g^{\alpha\beta}\Phi_{,\alpha;\beta}$ and some of the terms on the right-hand side are present in the energy-momentum tensor of (1.10). In fact, one can rewrite (1.13) for the static case as a Poisson-type equation with an effective source density for the Φ-field. In general (1.13) is a wave equation with a source induced by the fifth dimension.

Let us now leave the gauge for neutral matter (1.7) and focus on a special case of it, called the canonical gauge. This was the brainchild of Mashhoon, who realized that if one factorizes the 4D part of a 5D metric in a way which mimics the use of cosmic time in cosmology, significant simplification follows for both the field equations and especially the equations of motion (Mashhoon, Liu and Wesson 1994). The efficacy of this gauge is related to the fact that a quadratic factor in l on the 4D part of a 5D model has algebraic consequences similar to those of a quadratic factor in t on the 3D part of a 4D cosmological model. The latter case, in the context of 4D Friedmann-Robertson-Walker (FRW) cosmologies, is known as the Milne universe. This has several interesting properties (Rindler 1977). We will come back later to the Milne universe as a lower-dimensional example of highly-symmetric 5D manifolds. For now, we note the form of the metric and summarize its properties.

The 5D canonical metric has a line element given by

$$dS^2 = \frac{l^2}{L^2} g_{\alpha\beta}\left(x^\gamma, l\right) dx^\alpha dx^\beta - dl^2 \quad , \qquad (1.14)$$

where $x^4 = l$ is the extra coordinate and L is a constant length introduced for the consistency of physical dimensions. There is an extensive literature on (1.14), both with regard to solutions of the field equations (1.5) and the equations of motion which follow from minimizing the interval S in (1.14). Some of the consequences of (1.14) can be inferred from what we have already learned, while some will become apparent from later study. But for convenience we here summarize all of its main properties following Wesson (2002): (a) Mathematically (1.14) is general, insofar as the five available coordinate degrees of freedom have been used to set $g_{4\alpha} = 0$, $g_{44} = -1$. Physically, this removes the potentials of electromagnetic type and flattens the potential of scalar type. (b) The metric (1.14) has been extensively used in the field equations, and many solutions are known. These include solutions for the 1-body problem and cosmology which have acceptable dynamics and solutions with the opposite sign for g_{44} which describe waves. (c) When $\partial g_{\alpha\beta} / \partial l = 0$ in (1.14), the 15 field equations $R_{AB} = 0$ of (1.5) give back the Einstein equations as described above, now in the form $G_{\alpha\beta} = 3g_{\alpha\beta} / L^2$. These in general identify the scale L as the characteristic size of the 4-space. For the universe, the last-noted relations define an Einstein space with

$$\Lambda = 3/L^2 \quad , \tag{1.15}$$

which identifies the cosmological constant. (d) This kind of local embedding of a 4D Riemann space in a 5D Ricci-flat space can be applied to any N, and is guaranteed by Campbell's theorem. We will take this up in more detail below. (e) The factorization in (1.14) says in effect that the 4D part of the 5D interval is $(l/L)ds$, which defines a *momentum* space rather than a *coordinate* space if l is related to m, the rest mass of a particle. This has been discussed in the literature as a way of bridging the gap between the concepts of acceleration as used in general relativity, and force (or change of momentum) as used in quantum theory. (f) Partial confirmation of this comes from a study of the 5D geodesic and a comparison of the constants of the motion in 5D and 4D. In the Minkowski limit, the energy of a particle moving with velocity v is $E = l(1-v^2)^{-1/2}$ in 5D, which agrees with the expression in 4D if $l = m$. (g) The five components of the geodesic equation for (1.14) split naturally into four spacetime components and an extra component. For $\partial g_{\alpha\beta} / \partial l \neq 0$, the former contain terms parallel to the 4-velocity u^{α}, which do not exist in 4D general relativity. We will look into this situation later. But we note now that for $\partial g_{\alpha\beta} / \partial l = 0$, the motion is not only geodesic in 5D but geodesic in 4D, as usual. Indeed, for $\partial g_{\alpha\beta} / \partial l = 0$, we recover the 4D Weak Equivalence Principle as a kind of symmetry of the 5D metric.

The preceding list of consequences of the canonical metric (1.14) shows that it implies departures from general relativity when its 4D part depends on the extra coordinate, but inherits many of the properties of Einstein's theory when it does not. In the latter case, the 4D cosmological constant is inherited from the 5D scaling, and has a value $\Lambda = \pm 3 / L^2$ depending on the signature of the extra dimension $(\varepsilon = \mp 1)$. This is a neat result, and elucidates the use of de Sitter and anti-de Sitter spaces in approaches to cosmology and particle production, which use quantum-mechanical approaches such as tunneling. However, in general we might expect the potentials of spacetime to depend on the extra coordinate. Both for this case as in (1.14), and for the case where the scalar potential is significant as in (1.7), the vacuum will have a more complicated structure than that implied by the simple cosmological constant just noted. It was shown in (1.10) that in general the effective 4D energy-momentum tensor for neutral matter in 5D theory contains contributions from both ordinary matter and the vacuum. Ordinary matter (meaning material particles and electromagnetic fields) displays an enormous complexity of structure.

"Vacuum matter" (meaning the scalar field and virtual particles which defy Heisenberg's uncertainty relation) may display a corresponding complexity of structure. To use a cliché, 5D induced-matter theory implies that we may have only scratched the surface of "matter".

Membrane theory uses an exponential rather than the quadratic of (1.14) to factorize the 4D part of a 5D metric. Thus a generalized form of the type of metric considered by Randall and Sundrum (1998, 1999) is

$$dS^2 = e^{F(l)} g_{\alpha\beta} dx^\alpha dx^\beta - dl^2 \qquad . \qquad (1.16)$$

Here $F(l)$ is called the warp factor, and is commonly taken to depend on the cosmological constant Λ and the extra coordinate $x^4 = l$ in such a way as to weaken gravity away from the brane $(l = 0)$. Particle interactions, by comparison, are stronger by virtue of being confined to the brane, which is effectively the focus of spacetime. An important aspect of (at least) the early versions of brane theory is the assumption of Z_2 symmetry, which means in essence that the physics is symmetric about the hypersurface $l = 0$. This prescription is simple and effective, hence the popularity of membrane theory. However, a comparison of (1.16) and (1.7) shows that the former is merely a special case of the latter, modulo the imposition of the noted symmetry. In fact, examination shows that membrane theory and induced-matter theory are basically the same from a mathematical viewpoint, even if they differ in physical motivation. The most notable difference is that for membrane theory particles are confined to the spacetime hypersurface by the geometry, which is constructed with this in mind; whereas for induced-matter theory particles are only constrained by solutions of the 5D geodesic equation, and can wander away from spacetime at a slow rate governed by the cosmological constant or oscillate around it. That the field equations of membrane theory and space-time-matter theory are equivalent was shown by Ponce de Leon (2001). His work makes implicit use of embeddings, and we defer a discussion of these plus the connection between brane and STM worlds to the next section.

Embeddings must, however, play an important role in the extension of 5D theories to those of even higher dimension. That this is so becomes evident when we reflect on the preceding discussion. In it, we have morphed from 4D general relativity with Einstein's equations in the forms (1.1)–(1.4), to 5D relativity with the apparently empty field equations (1.5). These lead us to consider the electromagnetic gauge (1.6) and the gauge for neutral matter (1.7). The latter has associated with it the 5D Ricci components (1.8), which imply the 4D Ricci scalar (1.9) and the effective 4D energy-momentum tensor (1.10). The latter balances the Einstein equations, and leaves us with the 4 vector terms which satisfy (1.11) by virtue of (1.12), plus the 1 scalar wave equation (1.13). When the 5D metric has a 4D part which is factorized by a quadratic in the extra coordinate, we obtain the canonical metric (1.14), which leads us to view the cosmological constant (1.15) as a scale inherited from 5D. When alternatively the 5D metric has a 4D part which is factorized by an exponential in the extra coordinate, we obtain the warp metric (1.16), which leads us to view spacetime as a singular surface in 5D. All of these results are entrained – in the sense that they follow from the smooth embedding of 4D in 5D. Certain rules of differential geometry underly this embedding. The main one of these is a theorem of Campbell (1926), which was revitalized by Tavakol and coworkers, who pointed out that it also constrains the reduction from general relativity in 4D to models of gravity in 3D and 2D which may be more readily quantized (see Rippl, Romero and Tavakol 1995). It is not difficult to see how to extend the formalism outlined above for $N > 5$, so yielding theories of supersymmetry, supergravity, strings and beyond. But in so doing there is a danger of sinking into an algebraic morass. An appreciation of embedding theorems can help us avoid this and focus on the physics.

1.5   A Primer on Campbell's Theorem

Embedding theorems can be classified as local and global in nature. We are primarily concerned with the former because our field equations are local. (The distinction is relevant, because global theorems are more difficult to establish; and since they may involve boundary conditions, harder to satisfy.) There are several local embedding theorems which are pertinent to ND field theory, of which the main one is commonly attributed to Campbell (1926). He, however, only outlined a proof of the theorem in a pedantic if correct treatise on differential geometry. The theorem was studied and established by

Magaard (1963), resurrected as noted above by Rippl, Romero and Tavakol (1995), and applied comprehensively to gravitational theory by Seahra and Wesson (2003). The importance of Campbell's theorem is that it provides an algebraic method to proceed up or down the dimensionality ladder N of field theories like general relativity which are based on Riemannian geometry. Nowadays, it is possible to prove Campbell's theorem in short order using the lapse-and-shift technique of the ADM formalism. The latter also provides insight to the connection between different versions of 5D gravity, such as induced-matter and membrane theory. We will have reason to appeal to Campbell's theorem at different places in our studies of 5D field theory. In the present section, we wish to draw on results by Ponce de Leon (2001) and Seahra and Wesson (2003), to give an ultra-brief account of the subject.

Campbell's theorem in succinct form says: Any analytic Riemannian space $V_n(s,t)$ can be locally embedded in a Ricci-flat Riemannian space $V_{n+1}(s+1,t)$ or $V_{n+1}(s,t+1)$.

We are here using the convention that the "small" space has dimensionality n with coordinates running 0 to $n-1$, while the "large" space has dimensionality $n+1$ with coordinates running 0 to n. The total dimensionality is $N=1+n$, and the main focus is on $N=5$.

To establish the veracity of this theorem (in a heuristic fashion at least), and see its relevance (particularly to the theories considered in the preceding section), consider an arbitrary manifold Σ_n in a

Ricci-flat space V_{n+1}. The embedding can be visualized by drawing a line to represent Σ_n in a surface, the normal vector n^A to it satisfying $n \cdot n \equiv n^A n_A = \varepsilon = \pm 1$. If e_α^A form an appropriate basis and the extrinsic curvature of Σ_N is $K_{\alpha\beta}$, the ADM constraints read

$$G_{AB} n^A n^B = -\frac{1}{2}\left(\varepsilon R_\alpha^\alpha + K_{\alpha\beta} K^{\alpha\beta} - K^2\right) = 0$$

$$G_{AB} e_\alpha^A n^B = K_{\alpha;\beta}^\beta - K_{,\alpha} = 0 \qquad \qquad (1.17)$$

These relations provide $1 + n$ equations for the $2 \times n(n+1)/2$ quantities $g_{\alpha\beta}$, $K_{\alpha\beta}$. Given an arbitrary geometry $g_{\alpha\beta}$ for Σ_n, the constraints therefore form an under-determined system for $K_{\alpha\beta}$, so infinitely many embeddings are possible. This implies that the embedding of a system of 4D equations like (1.1)–(1.4) in a system of 5D equations like (1.5) is always possible.

This demonstration of Campbell's theorem can easily be extended to the case where V_{n+1} is a de Sitter space or anti-de Sitter space with an explicit cosmological constant, as in brane theory. Depending on the application, the remaining $n(n+1) - (n+1) = (n^2 - 1)$ degrees of freedom may be removed by imposing initial conditions on the geometry, physical conditions on the matter, or conditions on a boundary.

The last is relevant to membrane theory with the Z_2 symmetry. To see this, let us consider a fairly general line element with $dS^2 = g_{\alpha\beta}(x^\gamma, l) dx^\alpha dx^\beta + \varepsilon\, dl^2$ where $g_{\alpha\beta} = g_{\alpha\beta}(x^\gamma, +l)$ for $l \geq 0$ and $g_{\alpha\beta} = g_{\alpha\beta}(x^\gamma, -l)$ for $l \leq 0$ in the bulk (Ponce de Leon 2001). Non-gravitational fields are confined to the brane at $l = 0$, which is a singular surface. Let the energy-momentum in the brane be represented by $\delta(l) S_{AB}$ (where $S_{AB} n^A = 0$) and that in the bulk by T_{AB}. Then the field equations read $G_{AB} = \kappa \left[\delta(l) S_{AB} + T_{AB} \right]$ where κ is a 5D coupling constant. The extrinsic curvature discussed above changes across the brane by an amount $\Delta_{\alpha\beta} \equiv K_{\alpha\beta}(\Sigma_{l>0}) - K_{\alpha\beta}(\Sigma_{l<0})$ which is given by the Israel junction conditions. These imply

$$\Delta_{\alpha\beta} = -\kappa\left(S_{\alpha\beta} - \frac{1}{3} S g_{\alpha\beta} \right) \quad . \tag{1.18}$$

But the $l = 0$ plane is symmetric, so

$$K_{\alpha\beta}(\Sigma_{l>0}) = -K_{\alpha\beta}(\Sigma_{l<0}) = -\frac{\kappa}{2}\left(S_{\alpha\beta} - \frac{1}{3} S g_{\alpha\beta} \right) \quad . \tag{1.19}$$

This result can be used to evaluate the 4-tensor

$$P_{\alpha\beta} \equiv K_{\alpha\beta} - K g_{\alpha\beta} = -\frac{\kappa}{2} S_{\alpha\beta} \quad . \tag{1.20}$$

However, $P_{\alpha\beta}$ is actually identical to the 4-tensor $\left(g_{\alpha\beta,4} - g_{\alpha\beta}g^{\mu\nu}g_{\mu\nu,4}\right)/2\Phi$ of induced-matter theory, which we noted above in (1.12). It obeys the field equations $P^{\beta}_{\alpha;\beta} = 0$ of (1.11), which are a subset of $R_{AB} = 0$. That is, the conserved tensor $P_{\alpha\beta}$ of induced-matter theory is essentially the same as the total energy-momentum tensor in Z_2-symmetric brane theory. Other correspondences can be established in a similar fashion.

The preceding exercise confirms the inference that induced-matter theory and membrane theory share the same algebra, and helps us understand why matter in 4D can be understood as the consequence of geometry in 5D.

1.6   Conclusion

In this chapter, we have espoused the idea that extra dimensions provide a way to better understand known physics and open a path to new physics.

The template is Einstein's general relativity, which is based on a fusion of the primitive dimensions of space and time into 4D spacetime. The feasibility of extending this approach to 5D was shown in the 1920s by Kaluza and Klein, and if we discard their restrictive conditions of cylindricity and compactification we obtain a formalism which many researchers believe can in principle offer a means of unifying gravity with the forces of particle physics (Section 1.2). 5D is not only the simplest extension of general relativity, but is also commonly regarded as the low-energy limit of higher-N theories. Most work has been done on two versions of 5D relativity which are similar mathematically but different physically. Induced-matter (or space-time-matter) theory is the older version. It views 4D mass and energy as consequences of the extra dimension, so realizing the dream of Einstein and others that matter is a manifestation of geometry. Membrane theory is the newer version of 5D relativity. It views 4D spacetime as a hypersurface or brane embedded in a 5D bulk, where gravity effectively spreads out in all directions whereas the interactions of particles are confined to the brane and so stronger, as observed. These theories are popular because they allow of detailed calculations, something which is not always the case with well-motivated but more complicated theories for $N > 5$ (Section 1.3). The field equations of all theories in $N \geq 4$ dimensions have basically the same structure, and this is why we treated them together in Section 1.4. There we concentrated again on the case $N = 5$, paying particular attention to the equations which allow us to obtain the 15 components of the metric tensor. In the classical view, these are potentials, where the 10-4-1 grouping is related to the conventional split into gravitational, electromagnetic and scalar fields. In the quantum view, the corresponding particles are the spin-2 graviton, the spin-1 photon and the spin-0 scalaron. The extension of the metric and the field equations to $N > 5$ is obvious, in which case other particles come in. However, the extension of general relativity to $N > 4$ needs to be guided by embedding theorems. The main one of these dates again from the 1920s, when it was outlined by Campbell. The plausibility of Campbell's theorem can be shown in short order using modern techniques, as can the mathematical equivalence of induced-matter theory and membrane theory (Section 1.5). In summary, the contents of this chapter provide a basis for writing down the equations for $2 < N < \infty$ and deriving a wealth of physics.

In pursuing this goal, however, some fundamental questions arise. In studying (say) 5D relativity, we introduce an extra coordinate ($x^4 = l$), and an extra metric coefficient or potential (g_{44}). The two are related, and by analogy with proper distance in the ordinary 3D space of a curved 4D manifold we can define $\left| \int g_{44}\left(x^{\alpha},l\right)dl \right|^{1/2}$ as the "size" of the extra dimension. Even at this stage, two issues arise which need attention.

What is the nature of the fifth coordinate? Possible answers are as follows: (1) It is an algebraic abstraction. This is a conservative but sterile opinion. It implies that l figures in our calculations, but either does not appear in our final answer, or is incapable of physical interpretation once we arrive there. (2) It is related to mass. This is the view of induced-matter theory, where quantities like the density and pressure of a fluid composed of particles of rest mass m can be calculated as functions of l from the field equations. Closer inspection shows that for the special choice of gauge known as the pure-canonical metric, l and m are in fact the same thing. We will return to this possibility in later chapters, but here note that in this interpretation the scalar field of classical 5D relativity is related to the

Higgs (or mass-fixing) field of quantum theory. (3) It is a length perpendicular to a singular hypersurface. This is the view of membrane theory, where the hypersurface is spacetime. It is an acceptable opinion, and as we have remarked it automatically localizes the 4D world. But since we are made of particles and so confined to the hypersurface, our probes of the orthogonal direction have to involve quantities related to gravity, including masses.

The other issue which arises at the outset with 5D relativity concerns the size of the extra dimension, defined as above to include both the extra coordinate *and* its associated potential. This is a separate, if related, issue to what we discussed in the preceding paragraph. We should recall that even in 4D relativity, drastic physical effects can follow from the mathematical behaviour of the metric coefficients. (For example, near the horizon of an Einstein black hole in standard Schwarzschild coordinates, the time part of the metric shrinks to zero while the radial part diverges to infinity.) This issue is often presented as the question: Why do we not see the fifth dimension? Klein tried to answer this, as we have seen, by arguing that the extra dimension is compactified (or rolled up) to a microscopic size. So observing it would be like looking at a garden hose, which appears as a line from far away or as a tube from close up. Since distances are related to energies in particle experiments, we would only expect the finite size of the fifth dimension to be revealed in accelerators of powers beyond anything currently available. This is disappointing. But more cogently, and beside the fact that it leads to conflicts, many researchers view compactification in its original form as a scientific cop-out. The idea can be made more acceptable, if we assume that the universe evolves in such a way that the fifth dimension collapses as the spatial part expands. But even this is slightly suspect, and better alternatives exist. Thus for membrane theory the problem is avoided at the outset, by the construction of a 5D geometry in which the world is localized on a hypersurface. For induced-matter theory, particles are constrained with respect to the hypersurface we call spacetime by the 5D equations of motion. In the latter theory, modest excursions in the extra dimension are in fact all around us to see, in the form of matter.

Let us assume, for the purpose of going from philosophy to physics, that a fifth dimension may exist and that we wish to demonstrate it. We already know that 4D general relativity is an excellent theory, in that it is soundly based in logic and in good agreement with observation. We do not desire to tinker with the logic, but merely extend the scope of the theory. Our purpose, therefore, is to look for effects which might indicate that there is something bigger than spacetime.

References

Arkani-Hamed, N., Dimopoulos, S., Dvali, G. 1998, Phys. Lett. B429, 263.

Arkani-Hamed, N., Dimopoulos, S., Dvali, G. 1999, Phys. Rev. D59, 086004.

Campbell, J.E. 1926, A Course of Differential Geometry (Clarendon, Oxford).

Gubser, S.S., Lykken, J.D. 2004, Strings, Branes and Extra Dimensions (World Scientific, Singapore).

Halpern, P. 2004, The Great Beyond (Wiley, Hoboken).

Magaard, L. 1963, Ph.D. Thesis (Kiel).

Mashhoon, B., Liu, H., Wesson, P.S. 1994, Phys. Lett. B331, 305.

Ponce de Leon, J. 2001, Mod. Phys. Lett. A16, 2291.

Ponce de Leon, J., 2002, Int. J. Mod. Phys. 11, 1355.

Randall, L., Sundrum, R. 1998, Mod. Phys. Lett. A13, 2807.

Randall, L., Sundrum, R. 1999, Phys. Rev. Lett. 83, 4690.

Rindler, W. 1977, Essential Relativity (2nd. ed., Springer, Berlin).

Rippl, S., Romero, C., Tavakol, R. 1995, Class. Quant. Grav. 12, 2411.

Seahra, S.S., Wesson, P.S. 2003, Class. Quant. Grav. 20, 1321.

Szabo, R.J. 2004, An Introduction to String Theory and D-Brane Dynamics (World Scientific, Singapore).

Wesson, P.S., Ponce de Leon, J. 1992, J. Math. Phys. 33, 3883.

Wesson, P.S. 1999, Space-Time-Matter (World Scientific, Singapore).

Wesson, P.S. 2002, J. Math. Phys. 43, 2423.

2. THE BIG BANG REVISITED

"Seek, and you shall find" (Matthew, New Testament)

2.1   Introduction

The big bang in 4D general relativity is a singularity, through which the field equations cannot be integrated. While the standard big bang can be viewed as a birth event, its lack of computability is considered by many researchers to be a drawback. Hence the numerous attempts which have been made to avoid it. Some of these are quite innovative, and include the proposals that it involved a transition from negative to positive mass (Hoyle 1975), a quantum tunneling event (Vilenkin 1982) and matter production from Minkowski space (Wesson 1985). The extension of the manifold from 4 to 5 dimensions brings in new possibilities, which we will examine in what follows. The standard class of 5D cosmological models was found by Ponce de Leon (1988). He solved the 5D field equations (1.5), assuming that the 5D metric was separable, so that the conventional 4D Friedmann-Robertson-Walker (FRW) models were recovered on hypersurfaces where the extra coordinate was held fixed. But while they are appealing from the buddhistic view, in that they are flat and empty in 5D (though curved with matter in 4D), the Ponce de Leon models are not unique. So after contemplating the flat approach we will consider others, in which the 4D big bang can be viewed as a shock wave, a bounce and even a black hole in 5D.

34

These possibilities are all mathematically viable, and might be considered as too much of a richness in return for the modest act of extending the dimensionality from 4 to 5. However, these models have the redeeming feature of being analyzable: we can now calculate what happens for $t < 0$ as well as insisting on agreement with astrophysical data for $t > 0$ (using the standard identification of the time $t = 0$ with the big bang). This is in the tradition of physics. We prefer to work out the properties of the early universe, instead of being obliged to accept its creation as fiat.

2.2   Flat 5D Universes

The standard 5D cosmological models of Ponce de Leon (1988) have been much studied. They can be given a pictorial representation using a combination of algebra and computer work (Wesson and Seahra 2001; Seahra and Wesson 2002). Since this also allows us to gain insight to the nature of the 4D singularity, we follow this approach here.

The models are commonly written in coordinates $x^0 = t$, $x^{123} = r\theta\phi$ and $x^4 = l$ (we absorb the speed of light and the gravitational constant through a choice of units which renders them unity). The line element is given by

$$dS^2 = l^2 dt^2 - t^{2/\alpha} l^{2/(1-\alpha)} \left(dr^2 + r^2 d\Omega^2 \right) - \frac{\alpha^2 t^2}{(1-\alpha)^2} dl^2 \quad , \quad (2.1)$$

where $d\Omega^2 \equiv \left(d\theta^2 + \sin^2\theta\, d\phi^2\right)$. The dimensionless parameter α is related to the properties of matter.

The latter can be obtained using the technique outlined in Chapter 1, where we used Campbell's theorem to embed 4D general relativity with Einstein's equations $G_{\alpha\beta} = 8\pi T_{\alpha\beta}\,(\alpha\beta = 0,123)$ in an apparently empty 5D manifold with field equations $R_{AB} = 0\,(A, B = 0,123,4)$. Here the effective or induced energy-momentum tensor can be taken as that for a perfect fluid with density ρ and pressure p. Then the class of solutions (2.1) fixes these quantities via

$$8\pi\rho = \frac{3}{\alpha^2\tau^2}, \quad 8\pi p = \frac{2\alpha - 3}{\alpha^2\tau^2}, \tag{2.2}$$

where $\tau = lt$ is the proper time. The equation of state is $p = (2\alpha/3 - 1)\rho$. For $\alpha = 3/2$, the scale factor of (2.1) varies as $t^{2/3}$, the density and pressure of (2.2) are $\rho = 1/6\pi\tau^2$ with $p = 0$, and we have the standard $k = 0$ dust model for the late universe. For $\alpha = 2$, the scale factor varies as $t^{1/2}$, $\rho = 3/32\pi\tau^2 = 3p$ and we have the standard $k = 0$ radiation model for the early universe. Cases with $\alpha < 1$ describe models that expand faster than the standard FRW ones and have inflationary equations of state (see below). It should be noted that ρ and p of (2.2) refer to the total density and pressure, respectively. These could be split into multiple components, including visible matter, dark matter, and possible vacuum (scalar) fields. The last could include a contribution from a time-variable cosmological "constant" (Overduin 1999), of the type indicated by data on the dynamics of galaxies and the age of the universe.

Physically, the Ponce de Leon cosmologies are very acceptable. Mathematically, they are flat in three dimensions, curved in four dimensions, and flat in five dimensions. This means that (2.1) in coordinates (t, r, θ, ϕ, l) is equivalent to 5D Minkowski space in some other coordinates (T, R, θ, ϕ, L) with line element

$$dS^2 = dT^2 - \left(dR^2 + R^2 d\Omega^2\right) - dL^2 \quad . \tag{2.3}$$

This does not have a big bang, but the four-dimensional part of equation (2.1) <u>does</u> (the 4-geometry is singular for $t \to 0$). A situation similar to this occurs in general relativity with the Milne model. In many books this is presented as one of the FRW class with negative spatial curvature, but a fairly simple coordinate transformation makes the metric a 4D Minkowski space, and accordingly it is devoid of matter (Rindler 1977). The coordinate transformation between equation (2.1) and equation (2.3) for the corresponding five-dimensional case is not simple. It is given by

$$T(t,r,l) = \frac{\alpha}{2}\left[\left(1 + \frac{r^2}{\alpha^2}\right)t^{1/\alpha}l^{1/(1-\alpha)} - \frac{t^{(2\alpha-1)/\alpha}l^{(1-2\alpha)/(1-\alpha)}}{(1-2\alpha)}\right]$$

$$R(t,r,l) = rt^{1/\alpha}l^{1/(1-\alpha)}$$

$$L(t,r,l) = \frac{\alpha}{2}\left[\left(1 - \frac{r^2}{\alpha^2}\right)t^{1/\alpha}l^{1/(1-\alpha)} + \frac{t^{(2\alpha-1)/\alpha}l^{(1-2\alpha)/(1-\alpha)}}{(1-2\alpha)}\right] \quad . \tag{2.4}$$

We have made an extensive study of these relations, in order to better understand the nature of the big bang.

The 4D physics occurs in the FRW-like coordinates (t, r) of (2.1) on a hypersurface $l = l_0$ (the angular variables play no physical role and may be suppressed). The 4D models may, however, be regarded as embedded in a flat space with the coordinates (T, R, L) of (2.3) and viewed therefrom. With the help of (2.4), we can thus obtain pictures of the 4D models and study the structure of their singularities. This can be done for a range of the assignable parameters (α, l_0). As an aid to visualization, we let r and R run over positive and negative values so that the images are symmetric about $R = 0$ (if it is desired to have r, $R > 0$, then one of the symmetric halves may be deleted). To the same end, we add lines of constant t that intersect the $R = 0$ plane orthogonally (in a Euclidean sense) and lines of constant r that run parallel to the symmetry plane at $R = 0$. The models grow in ordinary space as they evolve in time. We present informative cases which are illustrated in the accompanying figures.

Model I ($\alpha = 3/2$, $l_0 = 1$), This is the standard $k = 0$ model for the late universe. By (2.1) it has a scale factor that varies as $t^{2/3}$, and by (2.2) it has $p = 0$. The shape is parabolic, and lines of constant r meet at a pointlike big bang at $T = R = L = 0$.

Model II ($\alpha = 1/30$, $l_0 = 1$). This is an inflationary $k = 0$ model for the very early universe. By (2.1) it has a scale factor that varies as t^{30}, and by (2.2) it has $(\rho + 3p) < 0$, so what is sometimes called the gravitational density is negative and powers a strong

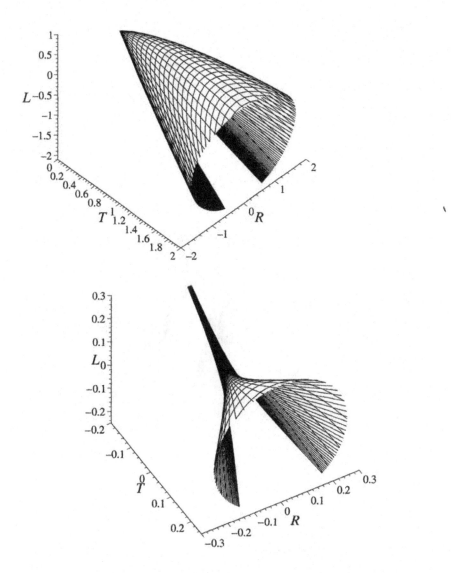

FIG. 2.1 – Hypersurfaces in 5D corresponding to spatially-flat FRW cosmologies in 4D. The upper case is the standard dust model, the lower case is an inflationary model, as discussed in the text.

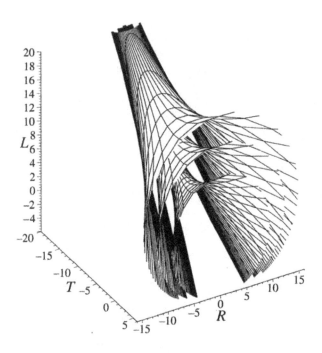

FIG. 2.2 – Hypersurfaces in 5D corresponding to spatially-flat, inflationary FRW cosmologies in 4D. The outer surface has $l_0 = 60$, the
middle surface has $l_0 = 40$ and the inner surface has $l_0 = 20$, as discussed in the text.

acceleration. Comoving trajectories converge to a point arrived at by following the null ray $T + L = R = 0$ into the past, towards past null infinity.

Model III ($\alpha = 1/3$, $l_0 = 20, 40, 60$). This is a set of inflationary $k = 0$ models with scale factors that vary as t^3, $(\rho + 3p) < 0$ and a common big bang at past null infinity.

Figures 2.1 and 2.2 reproduce known physics for the standard spatially-flat FRW models while adding a new perspective. Also, our figures for classical inflationary models are strikingly similar to those generated by computer for a stochastic theory of inflation based on quantum field theory (Linde 1994). In theories of the latter type, the rest masses of particles are basically zero and become finite through a mechanism involving the Higgs field (Linde 1990). In 5D classical theory, it has been argued that the Higgs potential is related to the g_{44}-component of the metric tensor (Wesson 1999). Alternatively, this component may be related to the effective 4D cosmological constant (Overduin 1999). In either case, we see from (2.1) that this factor is time-dependent, which raises the possibility of testing such models using particle masses and gravitational lensing.

In general, the dynamics of models like (2.1) may be studied by solving the 5D geodesic equation. We will give this detailed consideration elsewhere, but note here some results which follow if we use the 5D proper time S of (2.1) to characterize the motion. Thus the 5D geodesic gives the 5 velocities $U^A = dx^A / dS$ as $U^i = 0 (i = 1, 2, 3)$, with $U^0 = \mp \alpha (2\alpha - 1)^{-1/2} l^{-1}$ and $U^4 = \pm (1 - \alpha)^2 \alpha^{-1}$

$(2\alpha-1)^{-1/2}t^{-1}$. There is no motion in 3-space, and the galaxies are static with respect to each other because the coordinates (t, r, l) in (2.1) are designed to be spatially comoving. This is the same prescription as used in most presentations of the 4D Robertson-Walker metric (Rindler 1977). The motions detected spectroscopically by observational cosmology refer to a noncomoving frame. In five dimensions, the coordinate transformation to $\bar{t} = lt$, $\bar{r} = t^{1/\alpha}r$ and $\bar{l} = At^A l$ (where A is a constant introduced for the purely algebraic purpose of distinguishing $\bar{t}$ from $\bar{l}$) results in $\bar{U}^0 = \mp(2\alpha-1)^{1/2}\alpha^{-1}$, $\bar{U}^4 = 0$, $\bar{U}^\theta = \bar{U}^\phi = 0$ and $\bar{U}^1 = \mp(2\alpha-1)^{-1/2}\bar{r}/\bar{t}$. The last member is just Hubble's law.

We see that the 5D Ponce de Leon models (2.1) have the same law for galaxy motions as the standard 4D FRW models, as well as the same expressions for the density and pressure (2.2). However, the line element (2.1) may be connected to the 5D Minkowski one (2.3) by the coordinate transformations (2.4). This remarkable fact may be confirmed by computer, using a symbolic software package such as GRTensor (Lake 2004). There may be a singularity in the matter-filled and curved 4D space, but one does not exist in the empty and flat 5D space. In other words, the 4D big bang is due to an unfortunate choice of coordinates in a smooth 5D manifold. To this extent, it is something of an illusion.

2.3   The Singularity as a Shock Wave

In this section, we will summarize the properties of another exact solution of the 5D field equations whose Riemann-Christoffel tensor obeys $R_{ABCD} = 0$, meaning that it is flat. However, the new solution depends only on the combined variable $u \equiv (t - l)$, so it describes a wave. Solutions of this type in Newtonian hydrodynamics where the density ρ and/or pressure p change abruptly are called shock waves. We can evaluate these properties of matter as before, by studying the 4D Einstein equations $G_{\alpha\beta} = 8\pi T_{\alpha\beta}$ $(\alpha, \beta = 0,123)$ which are contained in the 5D field equations $R_{AB} = 0$ $(A, B = 0,123,4;$ see Wesson, Liu and Seahra 2000; Ponce de Leon 2003). The properties of the 5D solution imply that we can view the 4D singularity as a kind of shock wave.

The solution has a 5D line element given by

$$dS^2 = b^2 dt^2 - a^2 \left(dr^2 + r^2 d\Omega^2 \right) - b^2 dl^2$$

$$a = \left(hu \right)^{\frac{1}{(2+3\alpha)}}$$

$$b = \left(hu \right)^{-\frac{(1+3\alpha)}{2(2+3\alpha)}} \qquad . \tag{2.5}$$

The notation here is the same as in the preceding section, and the solution may be confirmed using the algebra of Chapter 1 or by computer (Lake 2004). It depends on 2 constants, h and α. The first has the physical dimensions of an inverse length or time, and is related to Hubble's parameter (see below). The second is dimensionless, and is related to the properties of matter. There is an associated equation of state, and after some algebra we find

$$p = \alpha\rho$$

$$8\pi\rho = \frac{3h^2}{\left(2+3\alpha\right)^2} a^{-3(1+\alpha)} \qquad . \qquad (2.6)$$

We see that $\alpha = 0$ corresponds to the late (dust) universe, and $\alpha = 1/3$ corresponds to the early (radiation) universe.

To elucidate the physical properties of the solution, it is instructive to change from the coordinate time t to the proper time T. This is defined by $dT = b\,dt$, so

$$T = \frac{2}{3}\left(\frac{2+3\alpha}{1+\alpha}\right)\frac{1}{h}(hu)^{\frac{3}{2}\left(\frac{1+\alpha}{2+3\alpha}\right)} \qquad . \qquad (2.7)$$

The 4D scale factor which determines the dynamics of the model by (2.5) and (2.7) is then

$$a(T) = \left[\frac{3}{2}\left(\frac{1+\alpha}{2+3\alpha}\right)hT\right]^{\frac{2}{3(1+\alpha)}} \qquad . \qquad (2.8)$$

For $\alpha = 0$, $a(T) \propto T^{2/3}$ as in the standard (Einstein-de Sitter) dust model. For $\alpha = 1/3$, $a(T) \propto T^{1/2}$ as in the standard radiation model. The value of Hubble's parameter is given by

$$H \equiv \frac{1}{a}\frac{\partial a}{\partial T} = \frac{1}{a}\frac{\partial a}{\partial t}\frac{dt}{dT} = \frac{h}{\left(2+3\alpha\right)}(hu)^{-\frac{3}{2}\left(\frac{1+\alpha}{2+3\alpha}\right)}$$

$$= \frac{2}{3(1+\alpha)T} \qquad . \qquad (2.9)$$

For $\alpha = 0$ and $1/3$, (2.9) shows that H has its standard values in terms of the proper time. We can also convert the density (2.6) from t to T using (2.7) and find

$$8\pi\rho = \frac{4}{3}\frac{1}{(1+\alpha)^2}\frac{1}{T^2} \quad . \tag{2.10}$$

For $\alpha = 0$ we have $\rho = 1/6\pi T^2$, and for $\alpha = 1/3$ we have $\rho = 3/32\pi T^2$, the standard FRW values. Thus, the 5D solution (2.5) contains 4D dynamics and 4D matter that are the same as in the standard 4D cosmologies for the late and early universe.

However, while the 5D approach does no violence to the 4D one, it adds significant insight. The big bang occurs in proper time at $T = 0$ by (2.10); but it occurs in coordinate time at $a = 0$ or $u = t - l = 0$ by (2.6) and (2.5). Now the field equations $R_{AB} = 0$ are fully covariant, so any choice of coordinates is valid. Therefore, we can interpret the physically-defined big bang either as a singularity in 4D or as a hypersurface $t = l$ that represents a plane wave propagating in 5D.

Some comments are in order about the shock-wave solution (2.5), the flat-universe solution (2.1) and certain other cosmological solutions in the literature (Wesson 1999). These all have $R_{ABCD} = 0$, and it is possible to make a systematic study of these equations (Abolghasem, Coley and McManus 1996). However, it is *not* usually possible to find coordinate transformations between solutions, or show the explicit transformation to 5D Minkowski space like (2.4), because of the level of complexity involved. Further, a given 5D solution may have different 4D interpretations. This is because the group of 5D coordinate transformations $x^A \rightarrow x^A \left(x^B \right)$ is wider that the 4D group $x^\alpha \rightarrow x^\alpha \left(x^\beta \right)$, so x^4-dependent transformations are mathematically equivalent in 5D but physically <u>non</u>-equivalent in 4D. The solutions in this and the preceding section provide an example of this. Both have $R_{ABCD} = 0$, but we interpret (2.1) as a 5D space with a 4D singularity embedded in it, and (2.5) as a wave moving in a 5D space which "pokes" through into 4D (like a 3D shock wave penetrates a 2D surface). The existence of multiple 4D interpretations of a given 5D solution raises an interesting question. If the real universe has one (or more) extra dimensions, then what coordinate system is being used for 4D cosmology? It seems to us that this question can be answered empirically, because the choice of coordinates in 5D affects the physics in 4D. An analogous situation occurs in the 4D/3D case and was touched on in the preceding section (see also Wesson 1999 pp. 100 – 102). In the 4D FRW models, the 3D spatial coordinates can be chosen as comoving so that the galaxies are fixed with respect to each other, or the coordinates can be chosen in such a way as to give the galaxies Hubble-law motions. In the 5D / 4D case, there is a similar situation which involves, among other things, the 3K microwave background. In the conventional 4D view, this is thermalized in the big-bang fireball. In the higher-dimensional view, some other mechanism must operate, such as a variation of particle masses that leads to efficient Thomson scattering (Hoyle 1975). We need to look into the detailed physics and decide by observational data which is the best approach.

2.4   A Bounce Instead of a Bang

Let us now move away from solutions which are flat to study one which is curved in 5D, and has the interesting property of a big "bounce" instead of a big bang in 4D.

It has frequently been speculated that 4D FRW models with positive 3D curvature might, after their expansion phase, recollapse to a big "crunch", from which they might re-emerge. However, this idea owes more to a belief in reincarnation than to physics, where it cannot be proved because it is impossible to integrate Einstein's equations through the second or (nth) singularity. It has also occasionally been suggested that an FRW model with a genuine bounce, where there is a contraction to a minimum but finite scale followed by an expansion, might serve to describe the real universe. But this idea fails when confronted with observational data, notably on the ages of globular clusters and the redshifts of quasars (Leonard and Lake; 1995; Overduin 1999). Furthermore, recent data on supernovae show that (provided there is negligible intergalactic dust) the universe on the largest scales is accelerating, implying a significant positive cosmological constant or some other dark form of energy with similar consequences (Perlmutter 2003). Indeed, the best fit on current data is to a universe which has approximately 70% of its density in the form of vacuum energy, approximately 30% in the form of dark but conventional matter, and only a smattering of luminous matter of the type we see in galaxies (Overduin and Wesson 2003). Thus realistic 5D models of the universe should contain a 4D cosmological "constant".

The latter, we saw in Chapter 1, is basically a measure of vacuum energy, and this can actually be variable. (In standard general relativity, the equation of state of the vacuum is $p_v = -\rho_v$ where $\rho_v = \Lambda / 8\pi$ in units where the speed of light and the gravitational constant are unity, with Λ a true constant.) Our goal, therefore, is to find a class of 5D solutions which not only replaces the singular big bang with a nonsingular bounce, but is rich enough to give back 4D matter of the appropriate normal and vacuum types.

A suitable class of models, which satisfies the 5D field equations $R_{AB} = 0$ and extends the 4D FRW ones, was studied by Liu and Wesson (2001; see also Liu and Mashhoon 1995). It has a line element given by

$$dS^2 = B^2 dt^2 - A^2 \left(\frac{dr^2}{1 - kr^2} + r^2 d\Omega^2 \right) - dl^2$$

$$A^2 = \left(\mu^2 + k \right) y^2 + 2vy + \frac{v^2 + K}{\mu^2 + k}$$

$$B = \frac{1}{\mu} \frac{\partial A}{\partial t} \equiv \frac{\dot{A}}{\mu} \quad . \tag{2.11}$$

Here $\mu = \mu(t)$ and $v = v(t)$ are arbitrary functions, k is the 3D curvature index ($k = \pm 1, 0$) and K is a constant. After a lengthy calculation, we find that the 5D Kretschmann invariant takes the form

$$I = R_{ABCD} R^{ABCD} = \frac{72K^2}{A^8} \quad , \quad (2.12)$$

which shows that K determines the curvature of the 5D manifold. From equations (2.11) we see that the form of $B\,dt$ is invariant under an arbitrary transformation $t = t(\bar{t})$. This gives us the freedom to fix one of the two arbitrary functions $\mu(t)$ and $v(t)$, without changing the basic solutions. The other arbitrary function can be seen to relate to the 4D properties of matter, which we now discuss.

The 4D line element is

$$ds^2 = g_{\alpha\beta} dx^\alpha dx^\beta = B^2 dt^2 - A^2 \left(\frac{dr^2}{1 - kr^2} + r^2 d\Omega^2 \right) . \quad (2.13)$$

This has the Robertson-Walker form which underlies the standard FRW models, and allows us to calculate the non-vanishing components of the 4D Ricci tensor:

$$^4R_0^0 = -\frac{3}{B^2} \left(\frac{\ddot{A}}{A} - \frac{\dot{A}\dot{B}}{AB} \right),$$

$$^4R_1^1 = {}^4R_2^2$$

$$= {}^4R_3^3 = -\frac{1}{B^2} \left[\frac{\ddot{A}}{A} + \frac{\dot{A}}{A} \left(\frac{2\dot{A}}{A} - \frac{\dot{B}}{B} \right) + 2k\frac{B^2}{A^2} \right] . \quad (2.14)$$

Now from (2.11) we have

$$B = \frac{\dot{A}}{\mu}, \qquad \dot{B} = \frac{\ddot{A}}{\mu} - \frac{\dot{A}}{\mu}\frac{\dot{\mu}}{\mu} \quad .$$

Using these in (2.14), we can eliminate B and $\dot{B}$ from them to give

$$^{4}R_{0}^{0} = -\frac{3\mu\dot{\mu}}{A\dot{A}}$$

$$^{4}R_{1}^{1} = {}^{4}R_{2}^{2} = {}^{4}R_{3}^{3} = -\left(\frac{\mu\dot{\mu}}{A\dot{A}} + \frac{2\left(\mu^{2} + k\right)}{A^{2}}\right) \qquad (2.15)$$

These yield the 4D Ricci scalar

$$^{4}R = -6\left(\frac{\mu\dot{\mu}}{A\dot{A}} + \frac{\mu^{2} + k}{A^{2}}\right) \qquad (2.16)$$

This, together with (2.15), enables us to form the 4D Einstein tensor $G_{\beta}^{\alpha} \equiv {}^{4}R_{\beta}^{\alpha} - \delta_{\beta}^{\alpha} \, {}^{4}R/2$. The nonvanishing components of this are

$$G_{0}^{0} = \frac{3\left(\mu^{2} + k\right)}{A^{2}}$$

$$G_{1}^{1} = G_{2}^{2} = G_{3}^{3} = \frac{2\mu\dot{\mu}}{A\dot{A}} + \frac{\mu^{2} + k}{A^{2}} \qquad (2.17)$$

These give the components of the induced energy-momentum tensor, since Einstein's equations $G_{\beta}^{\alpha} = 8\pi T_{\beta}^{\alpha}$ hold.

Let us suppose that the induced matter is a perfect fluid with density ρ and pressure p, moving with a 4-velocity $u^{\alpha} \equiv dx^{\alpha}/ds$, plus a cosmological term whose nature is to be determined. Then we have

$$G_{\alpha\beta} = 8\pi\left[\left(\rho + p\right)u_{\alpha}u_{\beta} + \left(\Lambda/8\pi - p\right)g_{\alpha\beta}\right] \qquad (2.18)$$

As in the FRW models, we can take the matter to be comoving in three dimensions, so $u^\alpha = (u^0, 0, 0, 0)$ and $u^0 u_0 = 1$. Then (2.18) and (2.17) yield

$$8\pi \rho + \Lambda = \frac{3\left(\mu^2 + k\right)}{A^2}$$

$$8\pi p - \Lambda = -\frac{2\mu\dot{\mu}}{A\dot{A}} - \frac{\mu^2 + k}{A^2} \quad . \tag{2.19}$$

These are the analogs for our solution (2.11) of the Friedmann equations for the FRW solutions. As there, we are free to choose an equation of state, which we take to be the isothermal one

$$p = \gamma\rho \quad . \tag{2.20}$$

Here γ is a constant, which for ordinary matter lies in the range (dust) $0 \leq \gamma \leq 1/3$ (radiation or ultrarelativistic particles). Using (2.20) in (2.19), we can isolate the density of matter and the cosmological term:

$$8\pi\rho = \frac{2}{1+\gamma}\left(\frac{\mu^2 + k}{A^2} - \frac{\mu\dot{\mu}}{A\dot{A}}\right)$$

$$\Lambda = \frac{2}{1+\gamma}\left[\left(\frac{1+3\gamma}{2}\right)\left(\frac{\mu^2 + k}{A^2}\right) + \frac{\mu\dot{\mu}}{A\dot{A}}\right] \quad . \tag{2.21}$$

In these relations, $\mu = \mu(t)$ is still arbitrary and $A = A(t,l)$ is given by (2.11). The matter density therefore has a wide range of forms, and the cosmological constant is in general variable.

Let us now consider singularities of the manifold in (2.11). Since this is 5D Ricci flat, we have $R = 0$ and $R^{AB}R_{AB} = 0$. The third

5D invariant is given by (2.12), from which we see that $A = 0$ (with $K \neq 0$) corresponds to a 5D singularity. This is a physical singularity, and as in general relativity, can be naturally explained as a big bang. However, from (2.16) and (2.21), we also see that if $\dot{A} = 0$, then all of the 4D quantities ^4R, ρ and Λ diverge. But while this defines a kind of 4D singularity, the 5D curvature invariant (2.12) does *not* diverge. This is a second kind of 4D singularity, associated with the minimum in the 3D scale factor A, and can be naturally explained as a big bounce. We further note from (2.11) that if $\dot{A} = 0$ then $B = 0$ (assuming $\mu \neq 0$), and the time part of the 4D line element vanishes. To sum up: the manifold (2.11) has a 5D geometrical singularity associated with $A = 0$ and a 4D matter singularity associated with $B = 0$, which we can explain respectively as a big bang and a big bounce.

The physics associated with the bounce, and plots of the 3D scale factor $A = A(t, l)$ as a function of the time t for various values of the extra coordinate l, were studied by Liu and Wesson (2001). They put $k = 0$ in (2.11) on the basis of observational data, and chose the functions $\mu(t)$ and $v(t)$ for algebraic convenience. They found that, typically, the form of the scale factor $A(t, l)$ is not symmetric around the minimum or time of the bounce. This implies that in 5D models of this type, there is a 4D production of entropy and/or matter around the bounce.

Extensive other work has been done on the class of solutions (2.11) due to the breadth of its algebra, with some interesting results for physics: (a) The big bounce has characteristics of an event horizon, at which the spatial scale factor and the mass density are finite, but where the pressure undergoes a sudden transition from negative to positive unbound values (Xu, Liu and Wang 2003). (b) The models are governed by equations which resemble the Friedmann relations for FRW cosmologies, but when the new solutions are compactified on an S_1 / Z_2 orbitold as in membrane theory, they yield 2 branes with different physical properties (Liu 2003). (c) By choosing the 2 arbitrary functions noted above in accordance with supernova data, the models before the bounce contract from a Λ-dominated vacuum, and after the bounce expand and indeed accelerate, with a dark energy contribution which is 2/3 of the total energy density for late times, in agreement with observations (Wang, Liu and Xu 2004). (d) This asymptotic behaviour can be shown by the use of a dynamical-systems approach to be universal, and due to the existence of two phase-plane attractors, one for the visible / dark-matter component and one for the scalar / dark-energy component (Chang et al. 2005). (e) The class of solutions (2.11) can, as mentioned before, be interpreted from the viewpoint of membrane theory, when the tension of the brane as a hypersurface in 5D and the strength of conventional gravity in 4D are constants (Ponce de Leon 2002). (f) The imposition of the Z_2 symmetry of membrane theory on the solutions (2.11) results in metrics which are even functions of the extra coordinate l, and when the dependency is via l^2 the bounce has the properties of a 4D phase transition (Liko and Wesson 2005). This compendium of properties does not exhaust the implications of the class (2.11). We have seen that we can interpret it as a bounce in a classical cosmology, a braneworld model, or a phase transition (which could be the classical analog of a discontinuity in a scalar Higgs-type quantum field). However, we will see in the following section that there is at least one more application of (2.11) that, while of a different kind, is just as remarkable.

2.5   The Universe as a 5D Black Hole

The concept of a 4D black hole is now so familiar that it is automatically associated with a central singularity, surrounded by an event horizon which depends on the mass M at the centre of spherically-symmetric 3D space, which latter is asymptotically flat. The latter property and others imply that the Schwarzschild solution is – up to coordinate transformations – unique. This is embodied in Birkhoff's theorem, which plays a significant role in Einstein's general theory of relativity. By comparison, the concept of a 5D "black hole" is considerably more complicated, due to the extra degrees of freedom introduced by the fifth coordinate. Solutions of the 5D field equations with a spherically-symmetric 3D space are called solitons. But even in the static case there is a class of solutions rather than a single one, and time-dependent cases are known (Wesson 1999). Thus Birkhoff's theorem, in its conventional form, fails in 5D. Indeed, it is unwise to carry over preconceptions about "black holes" from 4D to 5D. We will see below that it is more advisable to consider a topological 5D black hole, defined by the symmetries of its metric, and work out the properties of these without presumptions. Specifically, our aim in this section is to consider a general 5D metric of the black-hole type, and show that it is isometric to that of the cosmologies treated in the previous section (Seahra and Wesson 2003, 2005; Fukui, Seahra and Wesson 2001). That is, we wish to ponder the possibility that the universe may be a 5D black hole.

As in other sections in this chapter, we let upper-case English letters run 0–4 and lower-case Greek letters run 0–3. We use time and spherical-polar spatial coordinates plus a length l, so $x^A = (t; r \, \theta \phi \, ; \, l)$. Then $d\Omega_2^2 \equiv d\theta^2 + \sin^2 \theta d\phi^2$ is the measure on a 2D spherical shell. The topological black-hole class of 5D solutions is given by

$$dS^2 = h dT^2 - h^{-1} dR^2 - R^2 d\Omega_3^2 \quad . \qquad (2.22)$$

Here T is the time and $h = h(R)$ is a function of the radius R, where the latter is defined so that when the 3-measure $d\Omega_3 = d\Omega_3(k)$ reduces to the 2-measure defined above, then $2\pi R$ is the circumference. The Kretschmann scalar for (2.22) is $72K^2 R^{-8}$, where the 5D curvature K depends on the mass M at the centre of the 3-geometry (Seahra and Wesson 2003). This scalar is the same as (2.12) for the cosmological manifold (2.11). This coincidence and other properties suggest to us that (2.11) and (2.22) are geometrically equivalent descriptions of the same situation in different coordinates, or are isometries.

To prove this, we need to show a coordinate transformation which takes us from (2.22) to (2.11) or the reverse. We proceed to give the result, noting that it may be confirmed by computer. The radial transformation is specified by

$$R^2 = \left(\mu^2 + k\right)l^2 + 2vl + \left(v^2 + K\right)\left(\mu^2 + k\right)^{-1} \quad , \quad (2.23)$$

where μ and v are the functions of t in (2.11). This is an unusual mapping, which may repay further investigation. The corresponding temporal transformation turns out to have different forms, depending on whether $k = 0$ or $k = \pm 1$. To present these forms, we introduce a dummy variable $u = u(t)$ and the function $v = v(t,l) \equiv k\left[\left(\mu^2 + k\right)l + v\right]$ $\left(kK\right)^{-1/2} \mu^{-1}$. Then for $k = 0$, ± 1 we have respectively:

$$T = \frac{1}{K} \int \left\{ \frac{v^2}{\mu^3} v' - \frac{v\left(v^2 + K\right)}{\mu^4} \mu' \right\} du + \frac{1}{K} \left\{ \frac{\mu^3 l^3}{3} + \mu v l^2 + \left[\frac{v^2 + K}{\mu} \right] l \right\}$$

$$T = \frac{1}{k} \int \left\{ \frac{v'}{\mu} - \left[\frac{v}{\mu^2 + k} \right] \mu' \right\} du + \frac{1}{k} \left\{ \mu l - \frac{K}{2\left(kK\right)^{1/2}} ln\left[\frac{1+v}{1-v} \right] \right\}. \quad (2.24)$$

In these, it is to be understood that the integrals are over t and that the integrands involve $v = v(u)$, $\mu = \mu(u)$ with $v' \equiv dv/du$, $\mu' \equiv d\mu/du$. It should also be noted that while (2.24) relates the time for the black hole (2.22) back to the time for the cosmology (2.11), there is a special case of (2.22) where $kK < 0$ and $k = \pm 1$. However, for this case there is no Killing-vector defined horizon, so this would correspond to a naked singularity, with a negative mass. We therefore bypass this special case as unphysical, and conclude that (2.23) and (2.24) are in general the coordinate transformations that take the metric (2.22) for a 5D topological black hole back to the metric (2.11) for a 5D FRW-like universe.

The isometry just shown invites further analysis based on a comparison with the usual 4D Schwarzschild solution. The latter is commonly presented in coordinates where there is an horizon (defined by the mass M) which splits the manifold into parts, the distinction being from the geometrical viewpoint somewhat artificial. This problem is frequently addressed by introducing Kruskal-Szekeres coordinates, which effectively remove the horizon and extend the geometry. We are naturally interested in seeing if this is possible for the metric constructed from (2.23) and (2.24). There are a large number of choices for the parameters involved in these relations, so let us focus on the case where $k = +1$, $K > 0$. This means that the 3D submanifold is spherical, so we have an ordinary as opposed to a topological black hole. Then it may be shown that for our case there are KS-type coordinates U, V which are related to the R, T coordinates of (2.23) and (2.24) by

$$U = \mp \left[h(\mathrm{R}) \right] e^{R/M} e^{-T/M} \left| (R - M)/(R + M) \right|^{1/2}$$

$$V = \pm e^{R/M} e^{T/M} \left| (R - M)/(R + M) \right|^{1/2} \quad . \tag{2.25}$$

In these coordinates, the metric for the black hole is

$$dS^2 = M^2 \left(1 + M^2 / R^2\right) e^{-2R/M} dU dV - R^2 d\Omega_3^2 \quad . \quad (2.26)$$

A detailed investigation of the extended geometry corresponding to (2.26), including Penrose-Carter diagrams, appears elsewhere (Seahra and Wesson 2003). This elucidates the nature in which the manifold is covered by the coordinates of the cosmological metric (2.11), to which we now return.

By (2.11), an observer unaware of the fifth dimension or confined to a hypersurface in it would experience a 4D universe with line element

$$ds^2 = \left(\frac{1}{\mu}\frac{\partial A}{\partial t}\right)^2 dt^2 - A^2 \left(\frac{dr^2}{\left(1 - kr^2\right)} + r^2 \left(d\theta^2 + \sin^2 \theta d\phi^2\right)\right) . (2.27)$$

We discussed several interpretations of this in Section 2.4 preceding, which follow from choosing the two arbitrary functions $\mu(t)$ and $v(t)$ and evaluating its associated matter. We noted that the main feature of the class of solutions (2.11) is a bounce, where the spatial scale factor of (2.27) goes through a minimum. Now that we know that the cosmological metric is isometric to a black hole, it is easier to see what is involved: The scale factor $A(t, l)$ when it passes through the minimum ($\partial A / \partial t = 0$) induces a singularity in the metric which is of the same type as with a conventional black hole ($g_{00} = 0$). But we argued before that this singularity is not geometrical, and indeed it is now clear that it is of the kind found at the event horizon of a black hole. However, an observer in the 4D manifold (2.27) would interpret $A(t)$ as the standard scale factor of an FRW model if $A(t, l)$ evolves to be independent of the fifth dimension (see above). He would then wrongly assume that the universe starts in the state with $A = 0$ as a big bang, whereas it actually evolves from the state with $\partial A / \partial t = 0$ which is a big bounce. In the case where the bounce is associated with matter production, as we also mentioned in Section 2.4, it is useful to introduce the mass $\mathcal{M}$ of the fluid out to radius r as it is defined by the density ρ, pressure p and the metric (2.27). This is given by the Misner-Sharp-Podurets mass function (Misner and Sharp 1964; Podurets 1964; Wesson 1986). For the uniform fluid of (2.27), the relevant relations are

$$\mathcal{M} = 4\pi A r^3 \left(\mu^2 + k \right) = \left(4\pi / 3 \right) r^3 A^3 \rho$$

$$\partial \mathcal{M} / \partial t = -4\pi r^3 A^2 \left(\partial A / \partial t \right) p \quad . \tag{2.28}$$

These allow of matter production ($\partial \mathcal{M} / \partial t > 0$) both before the bounce ($\partial A / \partial t < 0, p > 0$) and after it ($\partial A / \partial t > 0, p < 0$), at least on the basis of classical theory. However, a proper investigation of this would require quantum theory, which would also help clarify the status of other issues with these models, such as inflation. ,

In this section, we have suggested that an observer living in a universe with the 4D line element (2.27) might be unaware that it is part of a 5D model of the form (2.11), which is geometrically equivalent to the 5D black hole (2.22). The argument has been mainly mathematical in nature. From a philosophical viewpoint, the idea that the universe is a higher-dimensional black hole may be harder to accept.

2.6   Conclusion

The 5D field equations $R_{AB} = 0$ lead to startling new cosmologies. However, the 5D equations contain the 4D Einstein ones $G_{\alpha\beta} = 8\pi T_{\alpha\beta}$, so we are in the comfortable situation of keeping what we know while finding something new. In this chapter we have looked at four new cosmologies. The first is a universe which is flat and empty in 5D, but contains on hypersurfaces the standard FRW models which are curved and have matter in 4D. We learn that the big bang may be a kind of artifact produced by an unfortunate choice of coordinates. The second example is also 5D-flat, but in it the big bang may be interpreted as the result of a shock wave propagating in the extra dimension. Our third example is based on a rich class of solutions where in general there is a big bounce rather than a big bang. The bounce may be associated with a phase transition and the creation of matter, at least in the case where the Z_2 symmetry of membrane theory is imposed. This view is in agreement with our fourth example, where we find that the bounce has some of the properties of an event horizon. This leads us to suggest that the universe may resemble a 5D black hole.

It is difficult to assess the plausibility of these and related ideas. However, many researchers would argue that none is intrinsically less plausible than the big bang. The latter phrase was coined by

Hoyle, who used it in a derogatory sense. To him, it appeared daft to assume that all of the matter in the universe was created in an initial singularity. The steady-state model of Bondi, Gold and Hoyle sought to provide a more logical alternative. In it, the dilution of matter by the universal expansion was compensated by its continual creation. This was studied using a modified form of the Einstein field equations by Hoyle and Narlikar, and in a different though related context by Dirac. It is well known that the steady-state cosmology foundered in the face of observational data, but its demise did not mean the end of new attempts to account for the nature and origin of matter. Interest in alternative theories continued, even after it was shown by the singularity theorems that in general relativity an initial singularity was inevitable given certain assumptions about the material content of the universe. And herein lies the gist: we are not sure of the nature of the very early universe, and so cannot be sure about its origin.

Modern 5D relativity has to be viewed against the historical backdrop just outlined. We may not be clear yet as to whether this manifold is smooth or has a membrane, but it can be argued that the 5D approach to cosmology is superior to all of its 4D predecessors. The induced-matter picture is particularly compelling. It uses the most basic mathematical object to form an exactly-determined set of field equations which describe not only the curvature of 4D spacetime but also its content of matter. And if we wish we can do away with the big bang.

References

Abolghasem, G., Coley, A.A., McManus, D.J. 1996, J. Math. Phys. 37, 361.

Chang, B., Liu, H., Liu, H., Xu, L. 2005, Mod. Phys. Lett. A 20, 923.

Hoyle, F. 1975, Astrophys. J. 196, 661.

Fukui, T., Seahra, S.S., Wesson, P.S., 2001, J. Math. Phys. 42, 5195.

Lake, K. 2004, GRTensor (Queen's U., Kingston).

Leonard, S., Lake, K. 1995, Astrophys. J. 441, L55.

Liko, T., Wesson, P.S. 2005, Int. J. Mod. Phys. A 20, 2037.

Linde, A.D. 1990, Inflation and Quantum Cosmology (Academic, Boston).

Linde, A.D. 1994, Sci Am. 271 (11), 48.

Liu, H., Mashhoon, B. 1995, Ann. Phys. 4, 565.

Liu, H., Wesson, P.S. 2001, Astrophys. J. 562, 1.

Liu, H. 2003. Phys. Lett. B 560, 149.

Misner, C.W., Sharp, D.H. 1964, Phys. Rev. B 136, 571.

Overduin, J.M. 1999, Astrophys. J. 517, L1.

Overduin, J.M., Wesson, P.S. 2003, Dark Sky – Dark Matter (Institute of Physics, Bristol).

Perlmutter, S. 2003, Phys. Today 56 (4), 53.

Podurets, M.A. 1964, Sov. Astron. (A.J.) 8, 19.

Ponce de Leon, J. 1988, Gen. Rel. Grav. 20, 539.

Ponce de Leon, J. 2002, Mod. Phys. Lett. A 17, 2425.

Ponce de Leon, J. 2003, Int. J. Mod. Phys. D 12, 1053.

Rindler, W. 1977, Essential Relativity (2nd ed., Springer, Berlin).

Seahra, S.S., Wesson, P.S. 2002, Class. Quant. Grav. 19, 1139.

Seahra, S.S., Wesson, P.S. 2003, J. Math. Phys. 44, 5664.

Seahra, S.S., Wesson, P.S. 2005, Gen. Rel. Grav., vol. 37, p.1339.

Vilenkin, A. 1982, Phys. Lett. B 117, 25.

Wang, B., Liu, H., Xu, L. 2004, Mod. Phys. Lett. A 19, 449.

Wesson, P.S. 1985, Astron. Astrophys. 151, 276.

Wesson, P.S. 1986, Phys. Rev. D 34, 3925.

Wesson, P.S. 1999, Space–Time–Matter (World Scientific, Singapore).

Wesson, P.S., Liu, H., Seahra, S.S. 2000, Astron. Astrophys. 358, 425.

Wesson, P.S., Seahra, S.S. 2001, Astrophys. J. 558, L75.

Xu. L., Liu, H., Wang, B. 2003, Chin. Phys. Lett. 20, 995.

3. PATHS IN HYPERSPACE

"Beam me up, Scottie" (Modern Startrek cliché)

3.1   Introduction

By the title of this chapter, it is implied that we will consider the possibility that a particle may move outside spacetime. In the early developmental stages of ND field theory, there was some discussion as to whether particles should move on the geodesics of familiar 4D space, or be allowed to wander into the higher dimensions. Our view is that if the fifth and higher dimensions are to be taken as "real" in some sense, then we should take the interval in the extended manifold, minimize it in analogy with Fermat's principle and other applications, and investigate the resulting dynamics. In this way, we can examine the acceptability of higher dimensions, and at least constrain them. Our view is that paths in $N > 4$D hyperspace are not the subject of theatrics, but rather provide a way of probing new physics.

There is, at the outset, an issue to be addressed which involves one of the long-standing differences between classical field theory and quantum mechanics. The equations of motion in general relativity involve the concept of acceleration, whereas the dynamics of particle physics uses the concept of momentum. The former concept involves only the 4D measures of space and time. The latter concept involves these plus the measure of mass. Of course, the two approaches overlap, and are indeed equivalent, in the case where the rest mass of an object is constant. However, there is a theoretical difference which goes to the root of what we mean by the concept of mass (Wesson 1999; Jammer 2000). And there is a practical difference, as can be appreciated in cases where the mass changes rapidly, as when a rocket burns fuel and leaves the Earth or a particle gains mass from the Higgs field in the early universe (Rindler 1977; Linde 1990). It will turn out that there are situations in which we need to consider carefully what happens when an object changes its rest mass as it pursues a path through a higher-dimensional manifold. Any student who observes the high velocity that a model rocket obtains by dint of shedding a fraction of its mass, knows that the concept of momentum is paramount. To this extent, we will need to consider how to introduce rest mass into the interval of general relativity, in a way which is consistent with other parts of the theory including the field equations (Wesson 2003a), and in agreement with observations such as those of QSOs which show a remarkable degree of uniformity in spectroscopic properties related to particle mass (Tubbs and Wolfe 1980). Since we can retain the usual definition of force as the product of acceleration and mass, another way of phrasing our objective is that we wish to elucidate forces in higher dimensions.

3.2   Dynamics in Spacetime

While this subject is one which we may be forgiven for assuming that we understand, it is instructive to remind ourselves of how accelerations enter general relativity and momenta enter quantum theory.

In Einstein's theory, the small element by which two arbitrary points in spacetime are separated is given by

$$ds^2 = g_{\alpha\beta} dx^\alpha dx^\beta \quad (\alpha, \beta = 0,123) \quad . \tag{3.1}$$

It is usual to take the interval (s) and the coordinates ($x^\alpha = t$, xyz or similar) to be lengths, while the components of the metric tensor ($g_{\alpha\beta}$) or potentials are dimensionless. The $g_{\alpha\beta}$ are given by solutions of the field equations (1.1), which being tensor relations obey the Covariance Principle, which means that they are valid in any system of coordinates (gauge). A typical example of a metric coefficient is $GM / c^2 r$ for the gravitational field outside an object of mass M at distance r in 3D spherically-symmetric space. (In this section we use physical units for the speed of light c, the gravitational constant G and Planck's constant h.) The Geodesic Principle asserts that the path of a particle is obtained by minimizing the interval via

$$\delta \left[\int ds \right] = 0 \quad . \tag{3.2}$$

The geodesic or path has 4 components. These are given by the geodesic equation, which in its most useful form reads

$$\frac{d^2 x^\alpha}{ds^2} + \Gamma^\alpha_{\beta\gamma} \frac{dx^\beta}{ds} \frac{dx^\gamma}{ds} = 0 \quad . \tag{3.3}$$

Here $\Gamma^\alpha_{\beta\gamma}$ are the Christoffel symbols of the first kind, which depend on the first partial derivatives of the $g_{\alpha\beta}$ (x^γ). In this way, (3.3) gives back in the weak-field limit $\left(\left|g_{\alpha\beta}\right| \ll 1\right)$ the standard Newtonian acceleration $(GM / c^2 r^2)$ outside an object like the Sun. When this is multiplied by the mass of a test object (m) we obtain the gravitational force, which balanced against the centrifugal force associated with a circular velocity (v) gives $GMm / r^2 = mv^2 / r$. In this, the masses on the left-hand side are actually gravitational in nature, while that on the right-hand side is inertial in nature (Jammer 2000). We have used the same symbol, because the Weak Equivalence Principle – as one of the founding bases of general relativity – asserts that they are essentially the same (see below). Then we can cancel the m on either side, obtaining the acceleration. The fact that this cancellation occurs is commonplace but of tremendous significance. It makes gravity a particularly simple interaction compared to others. For this and other reasons, the Equivalence Principle has been much tested and continues to be so, as we will see below. Here, we note that (3.1) – (3.3) provide a theory of dynamics based on accelerations, not forces. This is also apparent from the alternative but more compact form of (3.3) given by

$$u^{\alpha}_{;\beta}u^{\beta} = 0 \quad . \tag{3.4}$$

Here $u^{\alpha} \equiv dx^{\alpha} / ds$ is the 4-velocity, and the semicolon denotes the covariant derivative which takes into account the departure of the spacetime (3.1) from flatness. The rest mass m of a test particle does not appear. Indeed, it is acknowledged that this quantity requires a wider rationale, such as would be provided by Mach's Principle (Rindler 1977; Wesson, Seahra and Liu 2002). This principle motivated Einstein, but it is widely believed that it is not properly incorporated into standard general relativity. Thus the Covariance, Geodesic and Equivalence Principles on their own lead to a theory where masses and matter are sources which curve spacetime, but in which the motion of a particle does not depend on the mass of the latter.

In quantum theory, the situation is different. Here attention is focused on momentum as the product of mass and velocity, plus its integral the energy. In fact, much of the physics of the microscopic world can be summed up in one simple relation between the energy (E), the momentum (p) and the mass (m):

$$E^2 - p^2 c^2 = m^2 c^4 \quad . \tag{3.5}$$

This relation is closely obeyed by real particles (Pospelov and Romalis 2004). It is based on dividing the line element by the squared element of the proper time (ds^2), to obtain a dynamical relation. Alternatively, it is based on the convention that the 4-velocities are normalized via $u^\alpha u_\alpha = 1$ or 0, depending on whether the particle is massive or massless. Multiplying this by m^2 means that the 4-momenta are normalized via $p^\alpha p_\alpha = m^2$. With the usual identifications $\left(E = mc^2 u^0 , \ p^{123} = mcu^{123} \right)$, this results in the standard relation noted above. However, this approach contains <u>no</u> information about the possible case in which the mass of an object varies along its path via $m = m(s)$, which as we will see can happen in extended theories of gravity. This shortcoming carries over from the particle to the wave picture. The latter is commonly derived by using the operators

$$E \rightarrow \left(\frac{hc}{i\phi}\right)\left(\frac{\partial\phi}{\partial x^0}\right), \quad p \rightarrow \left(\frac{h}{i\phi}\right)\left(\frac{\partial\phi}{\partial x}\right) \tag{3.6}$$

to write the energy and 3-momentum of a particle in terms of a wave function ϕ. This causes the standard energy relation to become the Klein-Gordon equation. For the flat spacetime of special relativity, this reads

$$\Box^2\phi + (c/h)^2 m^2\phi = 0$$

where

$$\Box^2\phi \equiv \frac{\eta^{\alpha\beta}\partial^2\phi}{\partial x^\alpha \partial x^\beta} \quad (\eta^{\alpha\beta} = +1,-1,-1,-1) \tag{3.7}$$

The non-relativistic limit of the flat-space Klein-Gordon equation is the Schrodinger equation (which is used for systems like the hydrogen atom), and its factorized form is the Dirac equation (which is used for particles like the electron). For the curved spacetime of general relativity, it is necessary to proceed in a manner that is covariant. The action is

$$A \equiv \int \frac{p_\alpha}{h} dx^\alpha = \int \frac{mcu_\alpha}{h} dx^\alpha = \int \frac{mc}{h} ds \quad, \tag{3.8}$$

which is of course quantized. The action can be used to form a wave function $\phi \equiv e^{iA}$. The first derivative of this yields

$$p_\alpha = \left(h / i\phi \right) \partial \phi / \partial x^\alpha \quad , \tag{3.9}$$

as before. The second derivative of the wave function needs to be taken covariantly, however, to account for the curvature of spacetime. Using a semicolon to denote this and a comma to denote the ordinary partial derivative, we define as usual $\Box^2 \phi \equiv g^{\alpha\beta} \phi_{,\alpha;\beta}$. Then the second derivative of ϕ when contracted with $g^{\alpha\beta}$ yields

$$\Box^2 \phi + \frac{\phi}{h^2} p^\alpha p_\alpha = \frac{i\phi}{h} g^{\alpha\beta} p_{\alpha;\beta} \quad . \tag{3.10}$$

The l.h.s. of this is real while the r.h.s. is imaginary. The former gives $\Box^2 \phi + \left(c / h \right)^2 m^2 \phi = 0$, which is the Klein-Gordon equation for a curved spacetime. The latter gives $p^\beta_{;\beta} = 0$, which is the conservation equation for the momenta. Both of these relations are standard in particle physics.

The considerations of the two preceding paragraphs show that in general relativity and quantum theory the logic of standard dynamics is incomplete. In fact, 4D mechanics is based largely on well-chosen conventions.

3.3   Fifth Force from Fifth Dimension

Coordinates as well as conventions affect dynamics. In 4D general relativity, we saw in Section 2.3 that the galaxies can be considered static in comoving coordinates, or moving in accordance with Hubble's law in the frame frequently used in observational cosmology.

In 5D relativity, we will see in this section that the choice of coordinates (or gauge) affects not only the form of the metric but also the dynamics which follows from it. However, it will transpire that in the canonical coordinates of Section 1.4, the equations of motion in 5D become quite transparent (Wesson et al. 1999). Then they can be couched in the form of the usual 4D geodesic, plus an extra acceleration or force (per unit mass) due to the fifth dimension.

In conventional 4D dynamics, it is often stated that the 4-velocity u^α and the 4-force are orthogonal as in $u^\alpha f_\alpha = 0$ (the summation convention is in effect and we absorb the fundamental constants in this section by a choice of units). It is certainly the case that conventional electrodynamics and fluid motions obey such a law. However, it is also apparent that if we use the same formalism to set up laws of physics in (say) 5D, the relation $u^A f_A = 0$ would result in $u^\alpha f_\alpha = - u^4 f_4 \neq 0$ and a consequent departure from the 4D conservation laws ($A,B = 0,123,4$ where the argument can obviously be extended to higher dimensions). The condition of orthogonality is built into relativity. Given an ND line element $dS^2 = g_{AB} dx^A dx^B$ in terms of a metric tensor and coordinates, the velocities $U^A = dx^A / dS$ for a non-null path perforce obey $1 = U^A U_A$ as a normalization condition, and $U^B D_B U^A = 0$ as the condition which minimizes S in terms of a covariant derivative D_B which takes into account the curvature of the manifold. The latter N equations, when contracted, define a kind of force (per unit mass) F^A, and result in the common relation

$$U_A F^A = 0 \quad . \tag{3.11}$$

We see that this orthogonality condition depends on basic assumptions to do with the validity of Riemannian geometry and group theory. As such, it is difficult to believe that it could be contravened as a basis for modern physics.

It then follows that if the space is (say) 5D and not 4D, perfect conservation in the whole space implies imperfect conservation in the subspace (see above: $u^\alpha f_\alpha = - u^4 f_4 \neq 0$). Since we know that 4D laws are closely obeyed, this implies that the dimensionality of the world can be tested by looking for small departures from 4D dynamics.

The N conditions $U^B D_B U^A = 0$ are the analog of the 4D relations (3.4). The ND geodesic equation is the analog of the 4D one (3.3). With appropriately-defined Christoffel symbols, it is

$$\frac{dU^A}{dS} + \Gamma^A_{BC} U^B U^C = 0 \quad \left(A, B, C = 0 ... N \right) \quad . \tag{3.12}$$

Solutions of this can be found once the $\Gamma^A_{BC} = \Gamma^A_{BC} \left(g^{DE} \right)$ are known from solutions of the field equations, which are commonly taken to be $R_{AB} = 0$ (see Section 1.4). However, solutions of either the geodesic equation or the field equations in practice require some assumptions about $g_{AB} = g_{AB}(x^C)$. There are N arbitrary coordinates x^C, so in 5D we can apply 5 conditions to g_{AB} without loss of generality. Normally, these would be chosen with regard to some physical situation. But here, we adopt a different approach aimed at dynamics. There have actually been numerous attempts at solving (3.12) in 5D (Wesson 1999). Here we choose to retain contact with modern theory by factorizing the 4D part of the space using $x^4 = l$ in a way analogous to the synchronous coordinate system of general relativity (this does not restrict generality if $g_{\alpha\beta}$ is allowed to depend on l as well as x^γ). We also use the 5 coordinate degrees of freedom to set $g_{4\alpha} = 0$, $g_{44} = -1$, which in the usual interpretation of Kaluza-Klein theory suppresses effects of the electromagnetic and scalar fields (see Chapter 1). The interval then takes the canonical form of (1.14):

$$dS^2 = \frac{l^2}{L^2} ds^2 - dl^2 \quad (5D) \quad . \tag{3.13}$$

Here L is a constant introduced for the consistency of physical dimensions, assuming that the 4D interval is a length, which is given by

$$ds^2 = g_{\alpha\beta}\left(x^\gamma, l\right) dx^\alpha dx^\beta \quad (4D) \quad . \tag{3.14}$$

In other words, the 5D space contains the 4D space as the hypersurface $l = l(x^\gamma)$, providing a well-defined local embedding.

The utility of (3.13) and (3.14) becomes apparent when substituted into (3.12). We also note that since we make observations in 4D it is convenient to use s in place of S. The velocities are related by

$$U^A = \frac{dx^A}{dS} = \frac{dx^A}{ds}\frac{ds}{dS} = u^\alpha \frac{ds}{dS} \quad , \tag{3.15}$$

where by (3.13) we have

$$\frac{ds}{dS} = \left\{ \left(\frac{l}{L}\right)^2 - \left(\frac{dl}{ds}\right)^2 \right\}^{-\frac{1}{2}} \quad . \tag{3.16}$$

Then some algebra shows that (3.12) splits naturally into two parts. The 4D part reads

$$\frac{du^\alpha}{ds} + \Gamma^\alpha_{\beta\gamma} u^\beta u^\gamma = F^\alpha \quad , \tag{3.17}$$

whereas the part for the fifth dimension reads

$$\frac{d^2l}{ds^2} - \frac{2}{l}\left(\frac{dl}{ds}\right)^2 + \frac{l}{L^2} = -\frac{1}{2}\left[\left(\frac{l}{L}\right)^2 - \left(\frac{dl}{ds}\right)^2\right] u^\alpha u^\beta \frac{\partial g_{\alpha\beta}}{\partial l} \quad . \tag{3.18}$$

What we have done here is to split the dynamics, with the overlap confined to the extra acceleration or force (per unit mass) F^α. It is already obvious from (3.17) that conventional 4D geodesic motion is recovered if $F^\alpha = 0$, so let us consider this quantity.

The explicit form of F^α can be found by expanding the Γ^A_{BC} noted above. It is

$$F^\alpha = \left(-g^{\alpha\beta} + \tfrac{1}{2} u^\alpha u^\beta\right) \frac{dl}{ds} \frac{dx^\gamma}{ds} \frac{\partial g_{\beta\gamma}}{\partial l} \quad . \tag{3.19}$$

This is finite if the 4D metric depends on the extra coordinate and there is motion not only in 4D but also in the fifth dimension ($dl / ds \neq 0$). The latter is given by a solution of (3.18), and will in general be finite so the extra force (3.19) will also in general be finite.

Measuring (3.19) will require a combination of exact solutions and observations. But we expect the new force to be small because 4D dynamics is known to be in good agreement with available data (Will 1993). Also, F^α of (3.19) contains a part (N^α) normal to the 4-velocity u^α and a part (P^α) parallel to it, and strictly speaking it is the latter which violates the usual condition of orthogonality and would show the existence of an extra dimension. This part of the force (per unit mass) is given by

$$P^\alpha = \left(-\frac{1}{2}\frac{\partial g_{\beta\gamma}}{\partial l}u^\beta u^\gamma \right)\frac{dl}{ds}u^\alpha \quad , \tag{3.20}$$

or in short by $P^\alpha = \beta u^\alpha$ where β is a scalar that depends on the solution. There are numerous solutions known of the field equations $R_{AB} = 0$ (5D) which depend on $x^4 = l$ and therefore have finite β (Wesson 1999). It is also known that apparently empty 5D spaces can contain curved 4D subspaces of cosmological type with matter, as discussed in Chapter 1. It is even possible that the 5D space may be flat, as discussed in Chapter 2. We have therefore looked at a toy model where flat 5D space is written in Minkowski form. For this, $N^\alpha = 0$ and $P^\alpha = \beta u^\alpha$ with $\beta = (1/l)(dl/ds)$. Even this simple case has an extra acceleration, due essentially to the fact that we are using the 4D proper time s rather than the 5D interval S to parametize the dynamics. We will discuss astrophysical applications of the fifth force in detail in Chapter 5.

The form of the fifth force depends, as we have noted, on the form of the metric. Studies have been made for the induced-matter approach (Liu and Mashhoon 2000; Liu and Wesson 2000; Billyard and Sajko 2001; Wesson 2002a) and for the membrane approach (Youm 2000; Maartens 2000; Chamblin 2001; Ponce de Leon 2001). The results are conformable. However, the former when it uses the canonical metric (3.13) opens a unique physical vista to which we alluded in Section 1.4. There we learned that the constant L of the 5D metric can be identified from the field equations in terms of the 4D cosmological constant via $L^2 = 3/\Lambda$. Moreover, the first part of the 5D canonical line element (3.13) is identical to the action of 4D particle physics (squared) if the extra coordinate is identified in terms of the rest mass of a particle via $l = m$. Considerable work has been done on this intriguing possibility, to which we will return in Section 3.5. Clearly, if l is related to m then the rest masses of particles will in general vary with the 4D proper time by (3.18). However, for the FRW cosmologies this variation will be slow, because L is large as a consequence of Λ being small. It should also be mentioned that the variability of particle rest mass can be removed by using different parametizations for the dynamics. These include replacing the proper time by another affine parameter (Seahra and Wesson 2001), and replacing the geodesic approach by the Hamilton-Jacobi formalism (Ponce de Leon 2002, 2003, 2004). We realize that while the canonical metric (3.13) is convenient, it is also special in that it is *only* in this gauge that the extra coordinate l can be identified with the rest mass m as used in other applications of dynamics.

3.4   Null Paths and Two Times

Photons travel on paths in spacetime which are null with $ds^2 = 0$, and conventional causality is defined by $ds^2 \geq 0$. But the interval in an $N \geq 5D$ manifold need not necessarily be so restricted. A study of the 5D equations of motion suggests that particles with large charge/mass ratios can move on paths with $dS^2 < 0$ (Davidson and Owen 1986), and that particles with no electric charge can move on paths with $dS^2 = 0$ (Wesson 1999). The idea that massive particles on timelike geodesics in 4D are on null paths in 5D is in fact quite feasible, both for induced-matter theory (Seahra and Wesson 2001) and membrane theory (Youm 2001). Also, null paths are the natural ones for particles which move through fields which are solutions of the apparently empty Ricci-flat field equations.

The physics which follows from null paths can be different depending on the signature of the 5D metric. We kept open the sign of g_{44} in certain preceding relations (such as those for the electromagnetic and neutral-matter gauges in Section 1.4), but assumed that g_{44} was negative in some others (such as those for the canonical gauge). This is largely because astrophysical data indicate that the cosmological constant is positive, which for the induced-matter approach means that the last part of the metric has to be negative. However, much work on the quantum aspects of 5D relativity uses a de Sitter manifold with a negative cosmological constant (i.e. AdeS space), which would correspond to the opposite sign. Timelike extra dimensions are also used in certain models of string theory (e.g. Bars, Deliduman and

Minic 1999). It should be mentioned in this regard that the signature of the metric is important for finding solutions of the field equations, both for induced-matter and membrane theory. For example, the Ponce de Leon cosmologies considered in Section 2.2 exist only for $(+----)$. Conversely, the Billyard wave considered below exists only for $(+---+)$. This signature defines what is sometimes called a two-time metric. This may be a misleading name, insofar as the fifth dimension need not have the same nature as ordinary time. But in quantum theory, the statistical interaction of particles can actually lead to thermodynamic arrows of time for different parts of the universe which are different or even opposed (Schulman 2000). In general relativity, it is well known how to incorporate the phenomenological laws of thermodynamics; and of course a fundamental correspondence can be established between the two subjects via the properties of black holes. In higher-dimensional relativity, however, the situation is less clear, both for thermodynamics and mechanics. Indeed, some rather peculiar consequences follow for dynamics when we consider 5D two-time metrics Wesson (2002b). This is especially true when we couple this idea with that of null paths.

In the Minkowski gauge, a particle moving along a null path in a two-time 5D metric has

$$0 = dS^2 = dt^2 - (dx^2 + dy^2 + dz^2) + dl^2 \quad . \qquad (3.21)$$

The 5-velocities $U^A \equiv dx^A / d\lambda$, where λ is an affine parameter, obey $U^A U_A = 0$. With $\lambda = s$ for the proper 4D time, the velocity in ordinary space (v) is related to the velocity along the axis of ordinary time (u) and the velocity along the fifth dimension (w) by $v^2 = u^2 + w^2$. This implies superluminal speeds. But the particle which follows the path specified by (3.21) should not be identified with the tachyon of special relativity, because as we saw above, in 5D theory the extra coordinate $x^4 = l$ may not be an ordinary length. What we <u>can</u> infer, by analogy with 4D special relativity, is that all particles in the 5D manifold (3.21) are in causal contact with each other.

In 4D, causality is usually established by the exchange of light signals, which as viewed in a (3 + 1) split propagate as waves along paths with $ds^2 = 0$. While we do not know the nature of the corresponding mechanism in 5D with $dS^2 = 0$, it is instructive to consider two-time metrics with wave-like properties. There is one such solution of $R_{AB} = 0$ which has the canonical form (3.13) and is particularly simple (Billyard and Wesson 1996; Wesson 2001). It is given by

$$dS^2 = \frac{l^2}{L^2}\left[dt^2 - e^{i(\omega t + k_x x)}dx^2 - e^{i(\omega t + k_y y)}dy^2 - e^{i(\omega t + k_z z)}dz^2 \right] + dl^2 \ . \quad (3.22)$$

Here k_{xyz} are wave numbers and the frequency is constrained by the solution to be $\omega = \pm 2 / L$. We have studied (3.22) algebraically and computationally using the program GRTensor (which may also be used to verify it). Solution (3.22) has two "times". It also has complex metric coefficients for the ordinary 3D space, but closer inspection shows that the structure of the field equations leads to physical quantities that are real. The 3D wave is not of the sort found in general relativity, but owes its existence to the choice of coordinates. A trivial change in the latter suppresses the appearance of the wave in 3D space, in analogy to how a wave is noticed or not by an observer, depending on whether he is fixed in the laboratory frame or moving with the wave. A further change of coordinates can be shown to make (3.22) look like the 5D analog of the de Sitter solution. This leads us to conjecture that the wave is supported by the pressure and energy density of a vacuum with the equation of state found in general relativity, namely $p + \rho = 0$. This is confirmed to be the case, with $\Lambda < 0$. It may also be confirmed that (3.22) is not only Ricci-flat ($R_{AB} = 0$) but also Riemann-flat ($R_{ABCD} = 0$). It is a wave travelling in a curved 4D spacetime that is embedded in a <u>flat</u> 5D manifold which has no global energy.

As mentioned above, the logical condition on the path of a particle for such a background field is that it be null. Let us therefore consider a general case, where we take the metric not in the Minkowski form (3.21) but in the canonical form (3.22), thus:

$$0 = dS^2 = \frac{l^2}{L^2} ds^2 + dl^2 \quad . \tag{3.23}$$

Here we take $ds^2 = g_{\alpha\beta}(x^a, l) \, dx^\alpha dx^\beta$, using all of the 5 available coordinate degrees of coordinate freedom to suppress the potentials of electromagnetic and scalar type, but leaving the metric otherwise general. The solution of (3.23) is $l = l_0 \exp[\pm i \, (s - s_0) / L]$, where l_0 and s_0 are constants of which the latter may be absorbed. Then $l = l_0 \, e^{\pm is/L}$

describes an *l*-orbit which oscillates about spacetime with amplitude l_0 and wavelength L. The motion is actually simple harmonic, since $d^2l / ds^2 = -l / L^2$. Also $dl / ds = \pm il / L$, so the physical identification of the mass of a particle with the momentum in the extra dimension as in brane theory, or with the extra coordinate as in induced-matter theory are equivalent, modulo a constant. In both cases, the *l*-orbit may intersect the *s*-plane a large number of times. There is only one period in the metric (3.23), defined by L, but of course a Fourier sum of simple harmonics can be used to construct more complicated orbits in the l / s plane. [Alternatively, extra length scales can be introduced to (3.23) by generalizing L.] If we identify the orbit in the l / s plane with that of a particle, we have a realization of the old idea (often attributed to Wheeler and/or Feynman) that instead of there being 10^{80} particles in the visible universe there is in fact only one which appears 10^{80} times.

3.5   The Equivalence Principle as a Symmetry

The Weak Equivalence Principle is commonly taken to mean that in a gravitational field the acceleration of a test particle is independent of its properties, including its rest mass. This principle lies at the foundation of Einstein's theory of general relativity, and by implication is satisfied by other accounts of gravity which use 4D spacetime. In higher-dimensional theories, however, it is not clear if the principle holds. In 5D, there is an extra coordinate, which has an extra velocity or momentum associated with it. Both the extra coordinate and its rate of change have been linked in extensions of general relativity of the Kaluza-Klein type to the properties of a particle, such as its rest mass and electric charge. It would be facile to assume that all particles have the same values of those properties which can be attributed to the extra dimension. On the other hand, classic tests of the WEP have established its accuracy on the Earth to better than 1 part in 10^{12}. And new technology indicates that tests in space can push this to 1 part in 10^{18} or better (Lammerzahl, Everitt and Hehl 2001). In this section, therefore, we wish to collect previous results and give a coherent account of how the Equivalence Principle in its weak form relates to 5D gravity. (The strong form has not been so rigorously tested, since it implies that the laws of physics and their associated parameters are the same everywhere, including the remote parts of the universe.) Our aim is to constrain 5D relativity by the WEP, and better understand the nature of the latter.

We use the same terminology as before, with a 5D line element chosen to suppress electromagnetic effects but still algebraically general, that reads

$$dS^2 = g_{AB}\, dx^A\, dx^B = g_{\alpha\beta}\,(x^{\gamma}, l)\, dx^{\alpha}\, dx^{\beta} - \Phi^2\,(x^{\gamma}, l)\, dl^2 \quad . \qquad (3.24)$$

Here the 4D line element defines the conventional proper time via $ds^2 = g_{\alpha\beta}\, dx^{\alpha}\, dx^{\beta}$ with 4-velocities $u^{\alpha} \equiv dx^{\alpha}/ds$. As with previous studies of the fifth dimension, we prefer to use s rather than S as parameter because we wish to make contact with established physics. (This also allows us to handle the null 5D paths of Section 3.4 without difficulty.) With (3.24) as an algebraic basis and the WEP as a physical constraint, we now proceed to review certain subjects with a view to learning about the nature of the fifth coordinate.

(a) The extra force which appears when the manifold is extended from 4D to 5D has been derived in different ways for induced-matter theory and brane theory, and has been discussed in Section 3.3 in relation to the canonical metric (3.13). However, it also applies to the more general metric (3.24), and is in fact generic. To see this, we recall from Section 3.2 that there is a normalization condition on the 4-velocities:

$$g_{\alpha\beta}(x^{\gamma}, l)u^{\alpha}u^{\beta} = 1 \quad . \tag{3.25}$$

Let us consider a slight change in the 5D coordinates (including $x^4 = l$), by differentiating (3.25) with respect to s. Doing this and using symmetries under the exchange of α and β to introduce the Christoffel symbols $\Gamma^{\mu}_{\alpha\beta}$, there comes

$$2g_{\alpha\mu}u^{\alpha}\left(\frac{du^{\mu}}{ds} + \Gamma^{\mu}_{\beta\gamma}u^{\beta}u^{\gamma}\right) + \frac{\partial g_{\alpha\beta}}{\partial l}\frac{dl}{ds}u^{\alpha}u^{\beta} = 0 \quad . \tag{3.26}$$

This reveals that in addition to its usual 4D geodesic motion (the part inside the parenthesis), a particle feels a new acceleration (or force per unit mass). It is due to the motion of the 4D frame with respect to the fifth dimension, and is parallel to the 4-velocity u^{μ}. Explicitly, the parallel acceleration is

$$P^{\mu} = -\frac{1}{2}\left(\frac{\partial g_{\alpha\beta}}{\partial l}u^{\alpha}u^{\beta}\right)\frac{dl}{ds}u^{\mu} \quad . \tag{3.27}$$

This agrees with (3.20), which is the result of a longer if more informative derivation where the geodesic equation is applied to a canonical metric.   To return to the latter, let us write $g_{\alpha\beta}\left(x^{\gamma},l\right) = \left(l^2/L^2\right)\bar{g}_{\alpha\beta}\left(x^{\gamma}\right)$, where L is a length which by (1.15) is related to the 4D cosmological constant by $L^2 = 3/\Lambda$. The acceleration (3.27) can now be evaluated and simplified using (3.25).   Its nature becomes clear in the Minkowski limit, when the motion is given by

$$\frac{du^{\mu}}{ds} = P^{\mu} = -\frac{1}{l}\frac{dl}{ds}u^{\mu}$$

$$\text{or} \quad \frac{d}{ds}\left(lu^{\mu}\right) = 0 \quad . \tag{3.28}$$

This is just the expected law of conservation of linear momentum, provided $l = m$ is the rest mass of the particle.

(b) The action can be used to confirm this.   With coordinates such that $g_{\alpha\beta}\left(x^{\gamma},l\right) = \left(l^2/L^2\right)\bar{g}_{\alpha\beta}\left(x^{\gamma}\right)$ and $\Phi^2\left(x^{\gamma},l\right) = 1$, the 5D line element (3.24) is

$$dS^2 = \frac{l^2}{L^2}\bar{g}_{\alpha\beta}\left(x^{\gamma}\right)dx^{\alpha}dx^{\beta} - \Phi^2\left(x^{\gamma},l\right)dl^2 \quad . \tag{3.29}$$

This is the pure canonical form (3.13), for which by (3.19) the fifth force is zero and the motion is geodesic in the conventional 4D sense.

The first part of (3.29) gives back the element of action mds of particle physics provided $l = m$. We have mentioned this before, and recall two important things. Firstly, the rest mass of a particle in 5D theory may change along its path via $m = m(s)$, and even in 4D the action should properly be written as $\int mds$. Secondly, it is only in canonical coordinates that the simple identification $l = m$ can be made, though even in other coordinate systems the 4D action is part of a 5D one.

　　　(c) The 5D geodesic equation minimizes paths via $\delta\left[\int dS\right] = 0$, which generalizes the equations of motion in 4D and adds an extra component for the motion in the fifth dimension. This procedure can be carried out for the metric (3.24), which can be made even more general by including the electromagnetic potential. The working is long and boring. (See Wesson 1999, pp. 132 – 138 and pp. 161 – 167 for the cases where electromagnetism is and is not included respectively, as well as references to other work.) But the results of the noted variation can be summed up in terms of several fairly simple expressions, which under some circumstances are constants of the motion. Of these, let us consider the one associated with the zeroth or time component of the geodesic equation, which is normally associated with the particle energy when the metric is static. We take this, plus the assumptions that electromagnetic terms are absent, that the 3-velocity v is projected out, and that the 4-part of the metric is quadratically factorized in l as in the canonical case. Then the constant is

$$E = \frac{l}{\left(1-v^2\right)^{\frac{1}{2}}} \quad . \tag{3.30}$$

One does not have to be Einstein to see that this gives back the conventional 4D energy provided l is identified with the particle rest mass m.

(d) The field equations for 5D relativity are commonly taken to be $R_{AB} = 0$, which we learned in Chapter 1 contain by virtue of Campbell's theorem the 4D Einstein equations $G_{\alpha\beta} = 8\pi T_{\alpha\beta}$. The effective or induced energy-momentum tensor can be evaluated for the metric (3.24), in terms of quantities to do with the fifth dimension. It is given by (1.10) with the appropriate signature, namely:

$$8\pi T_{\alpha\beta} = \frac{\Phi_{,\alpha;\beta}}{\Phi} + \frac{1}{2\Phi^2}\left\{\frac{\Phi_{,4}g_{\alpha\beta,4}}{\Phi} - g_{\alpha\beta,44} + g^{\lambda\mu}g_{\alpha\lambda,4}g_{\beta\mu,4}\right.$$

$$\left. - \frac{g^{\mu\nu}g_{\mu\nu,4}g_{\alpha\beta,4}}{2} + \frac{g_{\alpha\beta}}{4}\left[g^{\mu\nu}_{,4}g_{\mu\nu,4} + \left(g^{\mu\nu}g_{\mu\nu,4}\right)^2\right]\right\} \quad . \tag{3.31}$$

This is known to give back the conventional matter content of a wide variety of 4D solutions, but in order to bolster the physical identification of l we note a generic property of it. For $g_{\alpha\beta,4} = 0$, (3.31) gives $8\pi T \equiv 8\pi g^{\alpha\beta}T_{\alpha\beta} = g^{\alpha\beta}\Phi_{,\alpha;\beta}/\Phi \equiv \Phi^{-1}\Box\Phi$. But the extra field equation $R_{44} = 0$, which we will examine below, gives $\Box\Phi = 0$ for

$g_{\alpha\beta,4} = 0$. Thus $T = 0$ for $g_{\alpha\beta,4} = 0$, meaning that the equation of state is that of radiation when the source consists of photons with zero rest mass. This is as expected.

(e) Algebraic arguments for $l = m$ can be understood from the physical perspective by simple dimensional analysis. The latter is actually an elementary group-theoretic technique based on the Pi theorem, and one could argue that a complete theory of mechanics ought to use a manifold in which spacetime is extended so as to properly take account of the <u>three</u> mechanical bases M, L, T. Obviously, this has to be done in a manner which does not violate the known laws of mechanics and recognizes their use of the three dimensional constants c, G and h. The canonical metric of induced-matter theory clearly satisfies these criteria, and we believe that it deserves its name because of the simplifications which follow from its quadratic l-factorization. But how unique is this form? To investigate, let us consider a 5D line element given by

$$dS^2 = \left(\frac{L}{l}\right)^{2a} \bar{g}_{\alpha\beta}\left(x^\gamma\right) dx^\alpha dx^\beta - \left(\frac{L}{l}\right)^{4b} dl^2 \quad . \quad (3.32)$$

Here a, b are constants which can be constrained by the full set of 5D field equations $R_{AB} = 0$. It turns out that there are 3 choices: $a = b = 0$ gives general relativity embedded in a flat and physically innocuous extra dimension; $a = -1$, $b = 0$ gives the pure canonical metric already discussed; while $a = b = 1$ gives a metric which looks different but is actually the canonical one after the coordinate transformation

$l \rightarrow L^2 / l$. We see that the last two cases describe the same physics but in terms of different choices of l. Temporarily introducing the relevant constants, these are

$$l_E = \frac{Gm}{c^2}, \quad l_P = \frac{h}{mc} \tag{3.33}$$

in what may be termed the Einstein and Planck gauges. These represent *convenient* choices of $x^4 = l$, insofar as they represent parametizations of the inertial rest mass m of a test particle which fit with known laws of 4D physics such as the conservation of linear momentum (see above: the fifth force conserves $l_E u^\mu$ or $l_P^{-1} u^\mu$). However, 5D relativity as based on the field equations $R_{AB} = 0$ is covariant under the 5D group of transformations $x^A \rightarrow \bar{x}^A(x^B)$, which is wider than the 4D group $x^\alpha \rightarrow \bar{x}^\alpha(x^\beta)$. Therefore 4D quantities $Q(x^\alpha, l)$ will in general change under a change of coordinates that includes l. This implies that we can only recognize m in certain gauges. The Einstein (canonical) and Planck gauges (3.33) are good parametizations for m, because they allow us to geometrize mass in a way consistent with the use of physical dimensions in the rest of physics.

The import of the preceding comments (a) – (e) is major for dynamics and the WEP, so a short recapitulation is in order. The general metric (3.24) is not factorized in the extra coordinate $x^4 = l$; but when the 4-velocities are normalized as in (3.25) there results the equation of motion (3.26) which includes a new acceleration parallel to the motion (3.27). This is identical to (3.20), which follows from the canonical metric (3.13), where the 4D part is factorized in l quadratically. This gauge conserves momentum via (3.28) if the identification $l = m$ is made. Other consequences follow for this gauge, as summarized in equations (3.29) – (3.33). It is apparent that the conventional concept of momentum is most conveniently realized by putting the 5D metric into the canonical form, where the extra coordinate plays the role of particle rest mass. Further, the Weak Equivalence Principle is then recovered from the equations of motion (3.17) – (3.19) when the 4D metric is independent of the extra coordinate. In other words, conventional dynamics and the WEP are the result of a metric *symmetry*.

This symmetry is geometrical in nature, but like the symmetries of particle physics, it can be expected to break down at some level. The identification $l = m$ suggested above means that our symmetry becomes mechanical, and that the breakdown would involve mass terms in the 4D part of the 5D metric. The WEP is usually phrased in terms of accelerations (rather than momenta), and is commonly understood to mean that the motion of an object of mass m in the gravitational field of a larger object of mass M is exactly geodesic in the 4D Einstein sense. A violation of the WEP would therefore follow from the presence of $l = m$ in the 4D part of the metric. Ideally, this would be formalized via a solution of the field equations, and we will consider such below. Here, however, we note that violations of the WEP are to be expected in situations where m / M is not negligible. Traditionally, such situations have been handled in areas like gravitational radiation by considering the "back reaction" of the test particle on the field of the source. But this is clearly an approximation to the real physics, which would involve the fields due to both objects. We are led to conclude that the WEP, viewed as a symmetry of 5D gravity, should be violated at some level.

3.6   Particle Masses and Vacua

In the foregoing section, we learned that in 5D relativity there is a convenient form of the metric which we renamed the Einstein gauge, because in it the extra coordinate essentially measures the mass of a particle by its Schwarzschild radius. However, there is an inverse form which we named the Planck gauge, because the correspondence involves the quantum of action, and effectively measures the mass of a particle by its Compton wavelength. The geometrization of rest mass in this manner is on a par with how Minkowski converted the time to a length using the speed of light, creating spacetime. We have merely extended this approach, creating a space-time-matter manifold.

In this section, we wish to address two issues which arise from the preceding account. They both concern the metric (3.24), which involves a scalar part and a spacetime part that can both depend on $x^4 = l$. Physically, these bring up questions of how to give definitions for the mass of a particle and the vacuum which are more general than those we considered above.

The mass of a particle in manifolds where there is a scalar field $|g_{44}| = \Phi^2(x^\gamma, l)$ can be defined most logically by

$$m \equiv \int |\Phi| \, dl = \int |\Phi(dl/ds)| \, ds \quad . \tag{3.34}$$

This is in line with how proper distance is defined in 3D (see Ma 1990). In practice, Φ would be given by a solution of the 5D field equations as outlined below, and dl/ds would be given by a solution of the extra component of the 5D geodesic equation (or directly from the metric for a null 5D path). We note that a potential problem with this approach is that Φ may show horizon-like behaviour. An example is the Gross/Perry/Davidson/Owen/Sorkin soliton, which in terms of a radial coordinate r which makes the 3D part of the metric isotropic has $g_{44} = -\Phi^2 = -[(1 - a/2r)/(1 + a/2r)]^{2\beta/\alpha}$ where a is the source strength and α, β are dimensionless constants constrained by the field equations to obey $\alpha^2 = \beta^2 + \beta + 1$ (see Wesson 1999, pp. 49 – 58). This problem may be avoided by restricting the physically-relevant size of the manifold. Another potential problem is that real particles may have $\Phi = \Phi(x^\gamma, l)$ so complicated as to preclude finding an exact solution. This problem may be avoided by expanding Φ in a Fourier series:

$$\Phi(x^\gamma, l) = \sum_{n=-\infty}^{+\infty} \Phi^{(n)}(x^\gamma) \exp(inl/L) \quad . \tag{3.35}$$

Here L is the characteristic size of the extra dimension, which is related to the radius of curvature of the embedded 4-space which the particle inhabits. It should be noted that in both modern versions of 5D relativity, namely induced-matter theory and brane theory, the extra dimension is not compactified. Thus we do not expect a simple tower of states based on the Plank mass, but a more complicated spectrum of masses that offers a way out of the hierarchy problem. This will depend on the precise form of $\Phi = \Phi(x^\gamma, l)$. To obtain this, we need to solve the $R_{44} = 0$ component of the field equations (1.13), which for the signature $(+ - - - -)$ being used here is

$$\Box \Phi = \frac{1}{2\Phi} \left[\frac{g^{\lambda\beta}_{,4} g_{\lambda\beta,4}}{2} + g^{\lambda\beta} g_{\lambda\beta,44} - \frac{\Phi_{,4} g^{\lambda\beta} g_{\lambda\beta,4}}{\Phi} \right] \quad . \quad (3.36)$$

This is a wave equation, and is source-free when the 4D part of the metric (3.24) does not depend on $x^4 = l$. But in general it will have such a dependency, and then (3.36) in combination with (3.34) raises the interesting possibility that the mass of a local particle depends on a universal scalar field. This is a realization of Mach's Principle, which we mentioned in Section 3.2 as a hypothesis for the origin of mass in classical field theory. There is also an obvious connection with the Higgs field, which is the agent by which particles obtain mass in quantum field theory.

The vacuum in manifolds where there is a spacetime $g_{\alpha\beta} = g_{\alpha\beta}(x^\gamma, l)$ is more complicated to define in 5D than it is in the 4D theory of Einstein. In the latter, there is a unique vacuum state which is measured by the cosmological constant and has the equation of state

$\rho_v = -p_v = \Lambda / 8\pi$. In 4D, the field equations $R_{\alpha\beta} = 0$ admit solutions which are empty of ordinary matter but have vacuum matter, the prime example being the de Sitter solution. In 5D, the field equations $R_{AB} = 0$ admit solutions which have ordinary matter and vacuum matter, in general mixed as evidenced by the effective energy-momentum tensor (3.31). The last relation, on inspection, shows that it is indeed more appropriate to talk of *vacua* in 5D, rather than *the vacuum* of 4D. To illustrate this, we note here 3 exact solutions of $R_{AB} = 0$:

$$dS^2 = \frac{l^2}{L^2}\left\{\left(1 - \frac{r^2}{L^2}\right)dt^2 - \frac{dr^2}{\left(1 - r^2/L^2\right)} - r^2 d\Omega^2\right\} - dl^2$$

$$dS^2 = \frac{l^2}{L^2}\left\{\left[\left(1 - \frac{r^2}{L^2}\right)^{1/2} + \frac{\alpha L}{l}\right]^2 dt^2 - \frac{dr^2}{\left(1 - r^2/L^2\right)} - r^2 d\Omega^2\right\} - dl^2$$

$$dS^2 = \frac{l^2}{L^2}\left\{\left[\left(1 - \frac{r^2}{L^2}\right)^{1/2} + \frac{\alpha L}{l}\right]^2 dt^2 - \frac{dr^2}{\left(1 - r^2/L^2\right)}\right.$$

$$\left. - \left(1 + \frac{\beta L^2}{rl}\right)^2 r^2 d\Omega^2\right\} - dl^2 \qquad . \tag{3.37}$$

Here $d\Omega^2 \equiv (d\theta^2 + \sin^2\theta d\phi^2)$, so all 3 solutions are spherically symmetric in 3D. The first is a 5D canonical embedding of the 4D de Sitter solution provided the identification $L^2 = 3 / \Lambda$ is made (see above). However, in general L measures the size of the potential well associated with $x^4 = l$. Solutions like these depend in general on two dimensionless constants α, β. We have examined the properties of (3.37) extensively, and have found that they are 5D flat ($R_{ABCD} = 0$; see Wesson 2003b). But here we note only their main 4D features. These can be appreciated by combining the above solutions in the form

$$dS^2 = \frac{l^2}{L^2}\left\{A^2 dt^2 - B^2 dr^2 - C^2 r^2 d\Omega^2\right\} - dl^2$$

$$A \equiv \left(1 - \frac{r^2}{L^2}\right)^{1/2} + \frac{\alpha L}{l}, \quad B \equiv \frac{1}{\left(1 - r^2/L^2\right)^{1/2}}, \quad C \equiv 1 + \frac{\beta L^2}{rl}. \quad (3.38)$$

The 4D subspaces defined by these solutions are curved, with a 4D Ricci scalar ^4R which by Einstein's equations is related to the trace of the 4D energy-momentum tensor by $^4R = -8\pi T$. The general expression for ^4R for any 5D metric of the form (3.24) is given by (1.9), which for the signature being used here is

$$^4R = -\frac{1}{4\Phi^2}\left[g^{\mu\nu}_{,4}g_{\mu\nu,4} + \left(g^{\mu\nu}g_{\mu\nu,4}\right)^2\right] \quad . \quad (3.39)$$

The special expression for (3.38) is

$$^4R = -8\pi T = -\frac{2}{L^2}\left[\frac{1}{AB} + \frac{2}{ABC} + \frac{1}{C^2} + \frac{2}{C}\right] \quad . \quad (3.40)$$

This shows that stress-energy is concentrated around singular shells where one of A, B or C is zero. The equation of state is in general anisotropic $\left(T_1^1 \neq T_2^2\right)$. If one replaces $1\,/\,L^2$ in (3.40) by its de Sitter limit $\Lambda\,/\,3$, it becomes obvious that the meaning of the cosmological "constant" requires a drastic rethink. The effective Λ is in general a function of r and l, opening the way to a resolution of the cosmological-constant problem.

3.7   Conclusion

Mechanics is often regarded as a staid subject, but its extension to $N \geq 5$ dimensions leads to some novel results. The 4D version is based on conventions, and in the extension to $N \geq 5$D we have to ensure that the dynamics is dimensionally consistent. The main conventions are the normalization of the velocities, and the definition of these in terms of the proper time (Section 3.2). The rest mass of a particle provides the link between the concept of acceleration as used in general relativity and the concept of momentum as used in quantum theory. If the manifold is extended to 5D in a meaningful way, it is inevitable that there appears an extra acceleration or force per unit mass (Section 3.3). This implies that our 4D laws are modified by the bits due to the fifth dimension. There are question marks over the size of the 5D interval and the sign of the fifth term in the metric (Section 3.4). It is possible that particles travelling on timelike paths in 4D are moving on null paths in 5D, so that massive particles in spacetime are like photons in the larger manifold. It is also possible that the fifth dimension is not spacelike as commonly assumed, but timelike, so particles can be multiply imaged. There is, though, no question that the Weak Equivalence Principle is observed to hold to great accuracy in 4D. This may be viewed as the consequence of a symmetry in 5D (Section 3.5). However, it would be simplistic to assume that the extra coordinate can never appear in the 4D part of the 5D metric. So like the symmetries of particle physics, it is expected that the WEP will be violated in some situations at some level. This is true for any interpretation of the fifth coordinate. When the extra coordinate is identified with particle rest mass – for which there are several arguments – the breakdown of the WEP is related to the ratio of the test and source masses. A more complete theory of particle mass should involve the scalar field (which acts like the classical version of the quantum Higgs field); but even when this is flat, a particle does not move through a unique vacuum but can experience one of several vacua (Section 3.6). Indeed, the multiple states of "emptiness" admitted by 5D relativity provide a fascinating topic for future study. If the extra coordinate is mechanical, the fact that it may in principle have either sign brings in the possibility of negative mass. And in general, exact vacuum solutions provide a new way to differentiate between 4D and 5D.

Questions to do with the nature of particle mass and the vacuum will be taken up below. It is clear that the results we have derived to here, in combination with others in the literature, bring us to the brink of quantum considerations. As a classical theory, it is apparent that 5D relativity is viable. We should recall that it agrees with the classical tests carried out in the solar system and with other astrophysical data, as well as being in conformity with less exact cosmological observations (Wesson 1999). It achieves this by virtue of being an extension of 4D general relativity rather than a departure from it. In fact, the only major uncertainty about 5D relativity is whether the manifold is smooth as for induced-matter theory or has a singular surface as for membrane theory. Unfortunately, the strength of the 5D approach is also its weakness, insofar as the departures it predicts from 4D theory are small and difficult to measure. Work is underway to quantify and detect classical effects which would indicate a fifth dimension, via for example a satellite test of the Equivalence Principle. Right now, however, we choose to leave the classical domain and turn to the quantum one.

References

Bars, I., Deliduman, C., Minic, D. 1999, Phys. Rev. D59, 125004.

Billyard, A., Wesson, P.S. 1996, Gen. Rel. Grav. 28, 129.

Billyard, A., Sajko, W.N. 2001, Gen. Rel. Grav. 33, 1929.

Chamblin, A. 2001, Class. Quant. Grav. 18, L17.

Davidson, A., Owen, D.A. 1986, Phys. Lett. B 177, 77.

Jammer, M. 2000, Concepts of Mass in Contemporary Physics and Philosophy (Princeton U. Press, Princeton).

Lammerzahl, C., Everitt, C.W.F., Hehl, F.W. 2001, Gyros, Clocks, Interferometers ... Testing Relativistic Gravity in Space (Springer, Berlin).

Linde, A.D. 1990, Inflation and Quantum Cosmology (Academic, Boston).

Liu, H., Mashhoon, B. 2000, Phys. Lett. A 272, 26.

Liu, H., Wesson, P.S. 2000, Gen. Rel. Grav. 32, 583.

Ma, G. 1990, Phys. Lett. A146, 375.

Maartens, R. 2000, Phys. Rev. D62, 084023.

Ponce de Leon, J. 2001, Phys. Lett. B 523, 311.

Ponce de Leon, J. 2002, Grav. Cosmol. 8, 272.

Ponce de Leon, J. 2003, Int. J. Mod. Phys. D12, 757.

Ponce de Leon, J. 2004, Gen. Rel. Grav. 36, 1335.

Pospelov, M., Romalis, M. 2004, Phys. Today 57 (7), 40.

Rindler, W. 1977, Essential Relativity (2nd. ed., Springer, Berlin).

Schulman, L.S. 2000, Phys. Rev. Lett. 85, 897.

Seahra, S.S., Wesson, P.S. 2001, Gen. Rel. Grav. 33, 1731.

Tubbs, A.D., Wolfe, A.M. 1980, Astrophys. J. 236, L105.

Wesson, P.S. 1999, Space-Time-Matter (World Scientific, Singapore).

Wesson, P.S., Mashhoon, B., Liu, H., Sajko, W.N. 1999, Phys. Lett. B 456, 34.

Wesson, P.S. 2001, Observatory 121, 82.

Wesson, P.S. 2002a, J. Math. Phys. 43, 2423.

Wesson, P.S. 2002b, Phys. Lett. B 538, 159.

Wesson, P.S., Seahra, S.S., Liu, H. 2002, Int. J. Mod. Phys. D11, 1347.

Wesson, P.S. 2003a, Int. J. Mod. Phys. D12, 1721.

Wesson, P.S. 2003b, Gen. Rel. Grav. 35, 307.

Will, C.M. 1993, Theory and Experiment in Gravitational Physics (Cambridge U. Press, Cambridge).

Youm, D. 2000, Phys. Rev. D62, 084002.

Youm, D. 2001, Mod. Phys. Lett. A16, 2371.

4. QUANTUM CONSEQUENCES

"To see a world in a grain of sand" (Blake)

4.1   Introduction

Practioners of quantum theory and classical theory often view the machinations of the other camp with suspicion. Certainly the subjects involve separate approaches and appear to have fundamental differences. Wave mechanics and modern quantum field theory depend on Planck's constant of action, and through it have an apparent level of non-predictability which is formalized in Heisenberg's uncertainty relation. General relativity and extensions of it depend on Newton's constant of gravity to measure the strength of a smooth field in which the motion of a test particle can be predicted with unlimited precision. It is frequently stated that quantum mechanics goes to classical mechanics in the limit in which Planck's constant tends to zero. But there is more to the issue than this, as can be appreciated by considering quantum and classical electrodynamics. The former uses the Dirac equation, which is first order and comes from a factorization of the metric of spacetime. The latter uses the Maxwell equations, which are second order and invariant under transformations of the whole spacetime. Both are highly successful theories, able to inform us respectively about (say) the spin of an electron or the emission of a radio wave. Their different qualities lie largely in their different algebras. Following from this, we will in this chapter study how 5D alge-

bra affects quantum and classical physics, with a view to their recon-
ciliation.

This goal may appear presumptuous. However, we will be
able to show significant technical results in key areas. These include
the inheritance of 4D Heisenberg-type dynamics from 5D laws of
motion, the plausibility that 4D mass is quantized because of the
structure of the fifth dimension, and the recovery of the 4D Klein-
Gordon and Dirac equations from 5D null paths. These topics, occu-
pying Sections 4.2 – 4.4, draw for their discussion on results we established in the preceding chapter. There, we saw that the extension
of the manifold from 4D to 5D necessarily brings in the existence of a
fifth force, raising the possibility that conservation laws in 5D are imperfect when viewed in 4D, and that we may observe as anomalous in
4D the bits of the dynamics left over from 5D. In the previous chapter we also saw that massive particles on timelike paths in 4D can be
viewed as moving on null or photon-like paths in 5D, implying that
objects which appear to be causally separated in 4D can be in contact
in 5D (like when objects apparently separated in ordinary 3D space
are connected by photons in 4D). These two properties of 5D relativity have, we should recall, been studied for both membrane theory
(Youm 2000, 2001) and induced-matter theory (Wesson 2002, 2003).
We will use the latter approach, since it lends itself more readily to
our purpose, but the two approaches are mathematically equivalent
(Ponce de Leon 2001). The null-path hypothesis can be applied in
two natural coordinate frames, which in the previous chapter we

dubbed the Einstein and Planck gauges. We will look at how quantization can depend on the gauge in Section 4.5, and discover how it may carry over from the microscopic to the macroscopic domain in the form of a (broken) symmetry for the spin angular momenta of gravitationally-dominated systems. Then in Section 4.6 we will employ the insights gained previously to revisit that most hoary of subjects, the difference between a particle and a wave. Encouragingly, we will find that the flexibility afforded by 5D coordinate transformations allows us to resolve this problem in terms of gauges. We round off our itinerary with a short discussion in Section 4.7, where we make some comments about unification.

4.2   4D Uncertainty from 5D Determinism

In this and the following section, to aid interpretation we use physical units for Planck's constant h, the gravitational constant G and speed of light c. We will also need to consider the cosmological constant Λ, which we take to have units of an inverse length squared. This parameter measures the energy density of the vacuum in 4D general relativity, and is related to the length L which scales the canonical metric in 5D theory, via $\Lambda = 3 / L^2$ (see Chapter 1). However, when the metric is transformed to suit other problems, L measures the scale of the potential well in which a particle finds itself, and may therefore be related to the vacuum or zero-point fields characteristic of quantum interactions.

The canonical line element which we have already examined has the form $dS^2 = (l/L)^2 ds^2 - dl^2$. The coordinates are $x^A = (x^\alpha, l)$ and the 5D interval contains the 4D one defined by $ds^2 = g_{\alpha\beta} dx^\alpha dx^\beta$ (α, β = 0, 123 for time and space). The 5D metric is mathematically general if we allow $g_{\alpha\beta} = g_{\alpha\beta} (x^\alpha, l)$, though it is physically special in that the electromagnetic potentials are $g_{4\alpha} = 0$ and the scalar potential is $g_{44} = -1$. However, we saw in Section 3.5 that if $g_{\alpha\beta} = g_{\alpha\beta} (x^\gamma$ only) then we recover the Weak Equivalence Principle as a geometrical symmetry. This and several other properties led us to identify the extra coordinate $x^4 = l = Gm/c^2$ in terms of the rest mass m of a test particle. Since this is the particle's Schwarzschild radius, we renamed the canonical metric the Einstein gauge. It has a coordinate transformation $l \to L^2/l$, which changes the form of the line element to one whose properties led us to identify the coordinate in the new metric as $l = h/mc$. Since this is the particle's Compton wavelength, we named this form the Planck gauge. It is specified by

$$dS^2 = \frac{L^2}{l^2} ds^2 - \frac{L^4}{l^4} dl^2 \quad , \qquad (4.1)$$

and it is this which we will study in the present section.

The dynamics associated with (4.1) are best brought out if we use the 4D proper time s as parameter, since we have a large body of data couched in terms of this, with which we wish to make contact.

The Lagrangian density $\mathcal{L} = (dS/ds)^2$ for (4.1) has associated with it 5-momenta given by

$$P_\alpha = \frac{\partial \mathcal{L}}{\partial(dx^\alpha/ds)} = \frac{2L^2}{l^2} g_{\alpha\beta} \frac{dx^\beta}{ds}$$

$$P_l = \frac{\partial \mathcal{L}}{\partial(dl/ds)} = -\frac{2L^4}{l^4} \frac{dl}{ds} \quad . \tag{4.2}$$

These define a 5D scalar which is the analog of the one used in 4D quantum mechanics:

$$\int P_A dx^A = \int \left(P_\alpha dx^\alpha + P_l dl \right)$$

$$= \int \frac{2L^2}{l^2} \left[1 - \left(\frac{L}{l} \frac{dl}{ds} \right)^2 \right] ds \quad . \tag{4.3}$$

This is zero for $dS^2 = 0$, since then (4.1) gives

$$l = l_0 e^{\pm s/L} \quad , \qquad \frac{dl}{ds} = \pm \frac{l}{L} \quad , \tag{4.4}$$

where l_0 is a constant. The second member of this shows why some workers have related the (inertial) rest mass of a particle to l (Wesson 2002) and some to its rate of change (Youm 2001) with consistent results: the two parametizations are essentially the same. In both cases, the variation is slow if $s/L \ll 1$ (see below). We prefer to proceed with the former approach, because it makes the first part of the 5D line element in (4.1) essentially the element of the usual 4D action $mc \, ds$, with the identification $l = h / mc$. It should be noted that the test particle we are considering has finite energy in 4D, but zero "energy" in 5D because $\int P_A dx^A = 0$.

The corresponding quantity in 4D is $\int p_\alpha dx^\alpha$, and using relations from the preceding paragraph it is given by

$$\int p_\alpha dx^\alpha = \int mu_\alpha dx^\alpha \int \frac{h \, ds}{cl} = \pm \frac{h}{c} \frac{L}{l} \qquad . \qquad (4.5)$$

The fact that this can be positive or negative goes back to (4.4), but since the motion is reversible we will suppress the sign in what follows for convenience. We will also put $L / l = n$, anticipating a physical interpretation which indicates that it is not only dimensionless but may be a rational number. Then (4.5) says

$$\int mc \, ds = nh \qquad . \qquad (4.6)$$

Thus the conventional action of particle physics in 4D follows from a null line element (4.1) in 5D.

The other scalar quantity that is of interest in this approach is $dp_\alpha \, dx^\alpha$. (It should be recalled that dx^α transforms as a tensor but x^α does not.) Following the same procedure as above, there comes

$$dp_\alpha dx^\alpha = \frac{h}{c} \left(\frac{du_\alpha}{ds} \frac{dx^\alpha}{ds} - \frac{1}{l} \frac{dl}{ds} \right) \frac{ds^2}{l} \qquad . \qquad (4.7)$$

The first term inside the parenthesis here is zero if the acceleration is zero or if the scalar product with the velocity is zero as in conventional 4D dynamics (see Section 3.3). But even so, there is a contribution from the second term inside the parenthesis which is due to the change in mass of the particle. This anomalous contribution has magnitude

$$\left|dp_\alpha dx^\alpha\right| = \frac{h}{c}\left|\frac{dl}{ds}\right|\frac{ds^2}{l^2} = \frac{h}{c}\frac{ds^2}{Ll} = n\frac{h}{c}\left(\frac{dl}{l}\right)^2 \quad , \qquad (4.8)$$

where we have used (4.4) and $n = L / l$. The latter implies $dn / n = -dl / l = dK_l / K_l$ where $K_l \equiv 1/l$ is the wavenumber for the extra dimension. Clearly (4.8) is a Heisenberg-type relation, and can be written as

$$\left|dp_\alpha dx^\alpha\right| = \frac{h}{c}\frac{dn^2}{n} \quad . \qquad (4.9)$$

This requires some interpretation, however. Looking back at the 5D line element (4.1), it is apparent that L is a length scale not only for the extra dimension but also for the 4D part of the manifold. (There may be other scales associated with the sources for the potentials that figure in $g_{\alpha\beta}$, and these may define a scale via the 4D Ricci scalar ^4R, but we expect that the 5D field equations will relate ^4R to L.) As the particle moves in spacetime, it therefore "feels" L, and this is reflected in the behaviour of its mass and momentum. Relations (4.6) and (4.9) quantify this. If the particle is viewed as a wave, its 4-momenta are defined by the de Broglie wavelengths and its mass is defined by the Compton wavelength. The relation $dS^2 = 0$ for (4.1) is equivalent to $P_A P^A = 0$ or $K_A K^A = 0$. The question then arises of whether the waves concerned are propagating in an open topology or trapped in a closed topology. In the former case, the wavelength is not constrained by the geometry, and low-mass particles can have large Compton wavelengths $l = h/mc$ with $l > L$ and $n = L/l < 1$. In the latter case, the wavelength cannot exceed the confining size of the geometry, and high-mass particles have small Compton wavelengths with $l \leq L$ and $n \geq 1$. By (4.9), the former case obeys the conventional uncertainty principle while the latter case violates it. This subject clearly needs an in-depth study, but with the approach adopted here we tentatively identify the former case as applying to real particles and the latter case as applying to virtual particles.

The fundamental mode ($n = 1$) deserves special comment. This can be studied using (4.6) – (4.9), or directly from (4.1) by using $l = h/mc$ with $dS^2 = 0$. The latter procedure gives $|dm| = m \, ds/L$ which with (4.6) yields $m = \left(\int mc \, ds \right)/cL = nh/cL$. This defines for $n = 1$ a fundamental unit mass, $m_0 = h/cL$. So apart from giving back a Heisenberg-type relation in (4.9), the 5D null-path approach appears to imply the existence of a quantum mass.

4.3   Is Mass Quantized?

This question cuts deeper than the formalism of any particular theory. It is basically asking if there is a minimum length scale for the universe, provided by as yet unproven properties of particle physics. If it exists, this small scale would compliment the large one which is indicated by the supernova and other data discussed in Chapters 1 and 2. There we learned that the cosmological constant is positive and finite, implying a scale for the present cosmos of order $\Lambda^{-1/2}$ or 10^{28} cm (Lineweaver 1998; Overduin 1999; Perlmutter 2003). We also learned that Λ could be interpreted as the energy density of the vacuum, which is $\Lambda c^4 / 8\pi G$ and constant in general relativity, but could be the result of a scalar field and time-variable in generalizations of that theory (Weinberg 1980; Wesson 1999; Padmanabhan 2003). It is plausible that any length scale – small or large – which we measure during our "snapshot" view of physics (of order 10^2 yr), could be changing over the longer periods typical of the evolution of the universe (of order 10^{10} yr). In this case the universe would be scale-free, or in the jargon of dimensional analysis, self-similar. It is hard to see how we could test this global hypothesis, attractive though it may be. But what we *can* do is take the generic approach provided by dimensional analysis, and see what it implies given present data about a smallest scale or a quantum of mass. We can then compare the result of this generic approach with the specific one provided by 5D relativity.

There are 2 masses which can be formed from a suite of 4 constants with degenerate or "overlapping" physical dimensions (Desloge 1984). Thus from h, G, c and Λ we can form a microscopic mass

$$m_P \equiv \left(\frac{h}{c}\right)\left(\frac{\Lambda}{3}\right)^{1/2} \simeq 2\times10^{-65}\,\mathrm{g} \quad , \tag{4.10}$$

and a macroscopic mass

$$m_E \equiv \left(\frac{c^2}{G}\right)\left(\frac{3}{\Lambda}\right)^{1/2} \simeq 1\times10^{56}\,\mathrm{g} \quad . \tag{4.11}$$

Here the two masses are relevant to quantum and gravitational situations, and so may be designated by the names Planck and Einstein respectively. [To avoid confusion, it can be mentioned that the mass $m_{PE} \equiv (hc/G)^{1/2} \simeq 5 \times 10^{-5}$ g which is sometimes called the Planck mass does *not* involve Λ and *mixes* h and G, so from the viewpoint of higher-dimensional field theory and (3.33), it arises from a mixture of gauges and is ill-defined, possibly explaining why this mass is not manifested in nature.] The mass (4.11) is straightforward to interpret: it is the mass of the observable part of the universe, equivalent to 10^{80} baryons of 10^{-24} g each. The mass (4.10) is more difficult to interpret: it appears to be the mass of a quantum perturbation in a spacetime with very small local curvature, measured by the astrophysical value of Λ as opposed to the one sometimes inferred from the vacuum fields of particle interactions (Weinberg 1980; Padmanabhan 2003). We will return to the cosmological-constant problem below, whose essence is that to make (4.10) and (4.11) the same would require $(hG / c^3) = 3 / \Lambda = L^2$, requiring the cosmological constant to have a value many orders bigger than what is inferred from astrophysics. This is a major problem, but is moot in the present context, because the astrophysical value of Λ is the smallest and so the mass given by (4.10) is also the smallest. That is, dimensional analysis predicts on general grounds a unit or quantum of mass of approximately 2×10^{-65} g.

Relativity in 5D provides a more specific means of analysis. As we have seen, there are two coordinate frames relevant to mechanical problems, with different identifications for the extra coordinate (3.33). These are $l_E = Gm / c^2$ and $l_P = h / mc$, for the Einstein (canonical) and Planck gauges respectively. These gauges have some common properties. For example, the null-path hypothesis results in $dl/ds = \pm l / L$ for both. This is relevant to the old version of Kaluza-Klein theory, in which l was compactified to a circle and its rate of change was related to electric charge, which was thereby quantized. However, we are using the extra dimension not as a means of understanding the electron charge but as a means of understanding mass scales, and we need to distinguish between the gauges. For this we use the labels introduced above. Thus

$$dS^2 = \left(\frac{L}{l_P}\right)^2 ds^2 - \left(\frac{L}{l_P}\right)^4 dl_P^2 \qquad (4.12)$$

is the Planck gauge which we used to study uncertainty in Section 4.2, and

$$dS^2 = \left(\frac{l_E}{L}\right)^2 ds^2 - dl_E^2 \qquad (4.13)$$

is the Einstein gauge which has been studied in various contexts as summarized elsewhere (Wesson 1999). These gauges imply different mass scales.

To see where the microscopic mass (4.10) originates, we can take (4.12) in the null case to obtain

$$\int d\left(\frac{L}{l_P}\right) = \left(\frac{1}{h}\right) \int mc\, ds \quad .$$
(4.14)

Here we know that the conventional action is quantized and equal to nh where n is an integer. Thus $L / l_P = n$. This means that the Compton wavelength of the particle cannot take on any value, but is restricted by the typical dimension of the (in general curved) spacetime in which it exists. Putting back the relevant parameters, the last relation says that $m = (nh/c)(\Lambda/3)^{1/2}$. For the ground state with $n = 1$, there is a minimum mass $(h/c)(\Lambda/3)^{1/2} \simeq 2 \times 10^{-65}$ g. This is the same as (4.10). To see where the macroscopic mass (4.11) originates, we can take (4.13) in the null case to obtain

$$\int \left(\frac{L}{l_E}\right) dl_E = \int ds \quad .$$
(4.15)

Here we do not have any evidence that the line element by itself is quantized, so the discreteness which is natural for the Planck gauge does not carry over to the Einstein gauge. However, in the Planck gauge the condition $L / l = n$ could have been used to reverse the argument and deduce the quantization of the action from the quantization of the fifth dimension, implying that the latter may be the fundamental assumption. Let us take this in the form $L/l_E = n$. (This implies $dl_E/ds = 1/n$, which holds too in the Planck gauge, so the velocity in the extra dimension is also quantized.) Putting back the relevant parameters, we obtain $m = (c^2/nG)(3/\Lambda)^{1/2}$. For the ground state with $n = 1$, there is a maximum mass $(c^2/G)(3/\Lambda)^{1/2} \simeq 1 \times 10^{56}$ g. This is the same as (4.11) above.

In summary, astrophysical data imply that we should add Λ to the suite of fundamental physical parameters, which implies on dimensional grounds that there are two basic mass scales related to h and G. These mass scales can be understood alternatively as the consequences of discreteness in the fifth dimension. The smaller scale defines a mass quantum of approximately 2×10^{-65} g. This is so tiny that mass appears in conventional experiments to be continuous.

4.4   The Klein-Gordon and Dirac Equations

We discussed the Klein-Gordon equation briefly in Section 3.2, noting that it is the wave equation which corresponds to the standard energy condition for a particle. Its low-energy form is the Schrodinger equation, and its factorized form is the Dirac equation. In the present section we wish to review the status of these equations in 4D and show how they can be derived from 5D (Wesson 2003). We use the terminology of previous sections in this chapter, except that we absorb the constants to smooth the algebra.

The Klein-Gordon equation relates the (inertial rest) mass m of a spin-0 particle to a scalar wave function ϕ via

$$\Box^2 \phi + m^2 \phi = 0 \quad . \tag{4.16}$$

Here $\Box^2 \phi \equiv \eta^{\alpha\beta} \partial^2 \phi / \partial x^\alpha \partial x^\beta$ for a flat 4D space $\eta^{\alpha\beta} = \text{diag}$ $(+1, -1, -1, -1)$ but the generalization to a curved 4D space $g^{\alpha\beta}$ is straightforward using the comma-goes-to-semicolon rule to introduce the covariant derivative, whence $\Box^2 \phi \equiv g^{\alpha\beta} \phi_{,\alpha;\beta}$. The 4-velocities $u^\alpha \equiv dx^\alpha / ds$ yield 4-momenta $p^\alpha \equiv mu^\alpha$ which can be obtained from a wave function via $p_\alpha = (1/i\phi) \partial \phi / \partial x^\alpha$ where $\phi = \exp\left[i \int p_\alpha dx^\alpha\right] =$ $\exp\left[i \int mds\right]$. The 4-velocities are conventionally normalized via $u^\alpha u_\alpha = 1$, which on multiplication by m^2 yields $p^\alpha p_\alpha = m^2$ or

$$E^2 - p^2 - m^2 = 0 \quad . \tag{4.17}$$

Here E and p are the energy and 3-momentum of the particle. These preliminaries may be familiar, but are necessary because it is required that (4.16) and (4.17) in 4D be recovered from a situation which is quite different in 5D. In such a manifold, the noted relations should clearly be replaced by

$$\Box^2 \Phi = 0 \tag{4.18}$$

$$P^A P_A = 0 \ , \tag{4.19}$$

where the 5D parameters are defined in analogous fashion to their 4D counterparts.

The embedding of 4D dynamics in 5D requires a choice of gauge, as mentioned above, and three such are in use. (a) The Minkowski gauge is flat 5D space, and with $x^4 = l_M$ the null path is given by $dS^2 = 0 = dt^2 - \left(dx^2 + dy^2 + dz^2 \right) - dl_M^2$, so $l_M = \pm s$ where a fiducial value of the 4D proper time has been absorbed. (b) The Planck gauge is (4.12) above, and with $x^4 = l_P$ the null path is given by $dS^2 = 0 = \left(L/l_P \right)^2 ds^2 - \left(L/l_P \right)^4 dl_P^2$, so $l_P = l_0 \exp\left(\pm s / L \right)$ where l_0 is a constant. (c) The Einstein gauge is (4.13) above, and with $x^4 = l_E$ the null path is given by $dS^2 = 0 = (l_E / L)^2 ds^2 - dl_E^2$, so $l_E = l_0 \exp\left(\pm s / L \right)$ as before. We noted previously that the Planck gauge is obtained from the Einstein gauge by the simple coordinate transformation $l_E \rightarrow L^2 / l_P$. In the Planck gauge, L is the scale of the potential well in which the particle moves. In the Einstein gauge this scale is large and related to the cosmological constant via $\Lambda = 3 / L^2$ of (1.15). The 3 gauges are suited to different kinds of problem. The Minkowski gauge is just a flat coordinate manifold; while the Planck and Einstein gauges with the identifications $l_P = 1/m$, $l_E = m$ of (3.33) are (in general) curved momentum manifolds. The connection between the null Minkowski and Einstein manifolds is via $l_E = l_0 \exp\left(\pm l_M / L \right)$, which with the coordinate transformation between the Einstein and Planck gauges noted above means that we can go between any of the three.

A wave function which is extended from 4D to 5D has the

form $\Phi = \exp\left[i \int\left(p_\alpha dx^\alpha + p_M dl_M\right)\right]$ in the Minkowski gauge.

Changing to the Einstein gauge, it transpires that the 4D and extra

parts of the integrand are equal, yielding $\Phi = \exp\left[2i \int mds\right]$. This

resembles the usual 4D wave function, but $m = m(s)$ in general so the

mass has to stay in the integrand. Also, it is algebraically more con-

venient to take the root of this quantity, so the wave function is just

$$\phi = \exp\left[i \int mds\right] \quad . \tag{4.20}$$

Taking the logarithm of this, writing dS_4 for the 4D part of the 5D

metric, and noting that for a null path $|dS_4| = dl_E$, there comes

$$ln(\phi) = i L \int|dS_4| = i L l_E \quad . \tag{4.21}$$

For a superposition of states, this invites us to form the natural loga-

rithm of the product of several wave functions, as

$$ln(\phi_1 \phi_2 \cdots) = i L\left(l_E^1 + l_E^2 + \cdots\right) \equiv i L l_E^{total} \quad . \tag{4.22}$$

That is, the wave function for a system of particles is basically the

total extension in the fifth dimension or the total mass.

Returning to (4.20), it gives

$$\frac{\partial^2 \phi}{\partial x^\alpha \partial x^\beta} = -\phi \left[m^2 \frac{ds}{dx^\alpha} \frac{ds}{dx^\beta} + i m \frac{du^\alpha}{ds} \left(\frac{ds}{dx^\alpha} \right)^2 \frac{ds}{dx^\beta} - i \frac{dm}{ds} \frac{ds}{dx^\alpha} \frac{ds}{dx^\beta} \right].$$

$$(4.23)$$

Here the last two terms in the bracket are related to the fifth force discussed in Section 3.3, and are imaginary. Taking the real part of (4.23) and contracting we obtain

$$g^{\alpha\beta} \frac{\partial^2 \phi}{\partial x^\alpha \partial x^\beta} = -m^2 \phi g^{\alpha\beta} v_\alpha v_\beta \quad , \qquad (4.24)$$

where $v_\alpha \equiv ds / dx^\alpha$, which implies $u^\alpha v_\alpha = 1$. However, the 4-velocities are normalized via $u^\alpha u_\alpha = 1$ (see above), so $v_\alpha = u_\alpha$. With the same definition of the d'Alembertian as before, (4.24) then reads

$$\Box^2 \phi + m^2 \phi = 0 \quad . \qquad (4.25)$$

This is formally the same as the conventional Klein-Gordon equation (4.16), but its physical interpretation is somewhat different insofar as $m = m(s)$ in general. In other words, 5D relativity yields the 4D Klein-Gordon equation but with a variable mass.

The Dirac equation relates the mass m of a spin-1/2 particle to a bispinor field ψ via

$$i\gamma^\alpha \frac{\partial \psi}{\partial x^\alpha} - m\psi = 0 \quad . \qquad (4.26)$$

Here γ^α are four 4×4 matrices which obey the relationship $\gamma^\alpha \gamma^\beta + \gamma^\beta \gamma^\alpha = 2\eta^{\alpha\beta}$. This decomposes the metric of spacetime, and

(4.26) is commonly regarded as resulting from the decomposition of the relativistic energy equation (4.17), or equivalently the relativistic wave equation (4.16). As above, these preliminaries may be familiar, but are necessary because it is required that (4.26) in 4D be recovered from a situation which is noticeably different in 5D. The main difference lies in the fact that the quantities which enter the problem in 4D are not covariantly defined whereas those in 5D are so. (The field ψ can be thought of as a 4-element column matrix, but neither this nor γ^α are 4-vectors.) This technical problem can be overcome by realizing that if ψ^* is the conjugate of ψ then the combination $\gamma^\alpha (\partial\psi/\partial x^\alpha)\psi^*$ must by (4.26) measure the real scalar quantity m, which can be identified as a geometrical quantity in 5D after choosing one of the gauges discussed above. A related issue concerns the distinction between the electron (e^-) and the positron (e^+). In 4D it is customary to regard the upper two elements of ψ as the spin states of the e^- and the lower two elements as the spin states of the e^+. But in 5D such an assignation cannot be made *a priori*, and the $e^\pm$ degeneracy is lifted by the presence of an external electromagnetic field (see below). A last issue which requires comment is the definition of an origin for the spin angular momentum of a system. In 4D, it is customary to set the zeroth component of the spin vector S_α to zero by referring the angular momentum to the centre of mass, the other 3 components being given by the relation $u^\alpha S_\alpha = 0$ which follows from the properties of the angular-momentum tensor. In 5D, a similar relation holds because S_A is spacelike while dx^A / dS is timelike, so their inner product can always be made to vanish. However, S_4 cannot in general be set to zero by an argument analogous to that for S_0, because insofar as angular momentum for the fifth dimension can be defined it involves a moment arm $x^4 = l$ which is related to particle mass. The upshot of these comments is that while it makes sense to seek a 5D analog of (4.26), certain differences of physical interpretation are to be expected.

Bearing this in mind, it is convenient to rewrite the spin / velocity orthogonality relation

$$u^A S_A = u^\alpha S_\alpha + u^4 S_4 = 0 \tag{4.27}$$

in the alternative form

$$u^\alpha S'_\alpha \pm w = 0 \quad , \tag{4.28}$$

where $S'_\alpha \equiv S_\alpha / S_4$ is defined in analogy with the electromagnetic potentials $A_\alpha \equiv g_{\alpha 4}/g_{44}$ of Kaluza-Klein theory and $w \equiv |u^4| = |dl / ds|$. These are related in the null Einstein gauge with electromagnetism (Section 1.4) by

$$dS^2 = 0 = \frac{l_E^2}{L^2} ds^2 - \left(dl_E + A_\alpha dx^\alpha \right)^2 \quad , \tag{4.29}$$

which gives

$$w = \frac{l_E}{L} \left| 1 + \frac{L}{l_E} A_\alpha u^\alpha \right| \quad . \tag{4.30}$$

Here the constant L can be absorbed within the Einstein gauge, and the weak-field limit of (4.30) with $l_E = m$ then causes (4.28) to read

$u^\alpha S'_\alpha \pm m = 0$. Redefining the scalar $u^\alpha S'_\alpha$ in terms of a bispinor field (see above) then gives a relation which is formally the same as (4.26). However, a more direct route to the Dirac equation than that provided by (4.27) is to *assume* the existence of a 5D field ψ (x^A) which obeys

$$i\gamma^A \frac{\partial \psi}{\partial x^A} = i\left(\gamma^\alpha \frac{\partial \psi}{\partial x^\alpha} + \gamma^4 \frac{\partial \psi}{\partial x^4} \right) = 0 \quad . \tag{4.31}$$

Here the Dirac matrices are extended to $\gamma^A (A = \alpha, 4)$; and $\partial \psi / \partial x^4 = (d\psi / ds)(ds / dx^4) = \pm (L / l)(d\psi / ds)$ for both the Einstein and Planck gauges. Within the latter, L can be absorbed and $l_P = 1/m$ causes (4.31) to become formally the same as (4.26), modulo γ^4. This has no obvious rationale, but may be given one via the Hoyle-Narlikar identity

$$\left| \bar{\psi}\psi \right|^2 \equiv \left(\bar{\psi}\gamma^\alpha \psi \right)\left(\bar{\psi}\gamma_\alpha \psi \right) - \left(\bar{\psi}\gamma_4 \psi \right)^2 \quad , \tag{4.32}$$

which may be proved using the Pauli matrices. We see that the 4D Dirac equation (4.26) can be understood as a consequence of the 5D relations (4.27) and (4.31), in the Einstein and Planck gauges, respectively.

4.5   Gauges and Spins

It is apparent from the contents of the preceding sections that quantization is, from a 5D standpoint, gauge-dependent. This should not be too surprising, if we recall that standard 4D relations to do with quantization cannot in general be invariant under a 5D change of gauge $x^A \rightarrow \bar{x}^A\left(x^\alpha, l\right)$ which involves the extra coordinate. However, it can then happen that a familiar quantization rule may take on a strange guise when the gauge is altered. In this section, we will briefly examine a situation of this kind, where the quantization of the spin in a microscopic state takes on a new appearance in a macroscopic state.

The basis for discussing spin is that Planck's constant h defines a unit for both the action <u>and</u> the spin angular momentum of a particle. (The latter involves a factor 2π, of course, but we are not here concerned with that.) It is reasonable to ask what happens to the spin when we change from the Planck gauge to the Einstein gauge. For both gauges, the null-path hypothesis with $n = L/l$ yields $ds = \pm n$ dl irrespective of the nature of l. Here dl corresponds to a change in n by $dn = -(L/l^2)\, dl$, so $ds = \pm l\, dn$ for both gauges. For the Planck gauge with $l = 1/m$, this means that $mds = \pm dn$. This is of course the standard rule for the action, and defines the quantum for it as h. This is also the quantum for the spin, insofar as bosons and fermions have integral and half-integral multiples of it. But if we use the previous relation $ds = \pm l\, dn$ with $l = m$ for the Einstein gauge, the implication is that it is now ds/m which takes on discrete values (rather than mds). The associated quantum has the physical dimensions of G/c^2 (rather than h). By analogy, we might expect a constant with these or similar dimensions to figure also in the spin angular momenta of gravitationally-dominated systems.

We can express the preceding argument in another way by restoring the physical constants. Then the Planck and Einstein gauges have quantization rules of the form

$$\int mcds = \pm nh \qquad (4.33)$$

$$\int \frac{cds}{m} = \pm np \quad , \qquad (4.34)$$

where p is a constant with the physical dimensions of $M^{-1} L^2 T^{-1}$, which by analogy with the dual role of h might be expected to be relevant to the spins of astrophysical systems.

In this regard, it is interesting to note that there is evidence of a kind of (broken) symmetry for the spin angular momenta (J) and masses (M) of astrophysical systems (Wesson 1981, 2005). It is expressed in the rule

$$J = p M^2 \quad , \qquad (4.35)$$

where the constant p has the required physical dimensions. (This symbol should not be confused with that for the 3-momentum of a particle used earlier.) The value of this constant may be determined from the extensive but older set of data illustrated in Figure 4.1, which agrees with the newer but more restricted set of data illustrated in Figure 4.2 (see Wesson 2005; Steinmetz and Navarro 1999; Bullock et al. 2001). The approximate value is $p \simeq 8 \times 10^{-16} g^{-1} cm^2 s^{-1}$. The relation (4.35) has been compared to the Regge trajectories of particle physics, but as a gravitational symmetry it is broken by forces

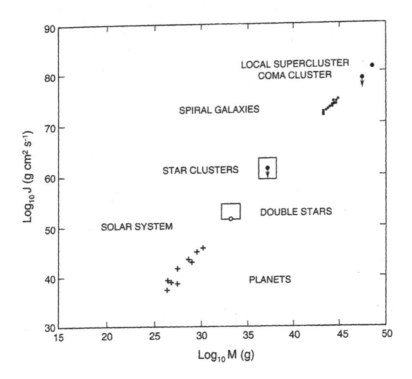

FIG. 4.1 – The angular momenta J of various astronomical systems versus their masses M. This plot is adapted from one by Wesson (1981). Data like those presented here have been much discussed in the literature, and what should be included / decluded remains controversial. (Asteroids are excluded here because they are not gravitationally dominated, being supported mainly by solid-state forces. Local supercluster dynamics are still under discussion.) The data support $J = pM^2$, where p is a constant whose value is $p \simeq 8 \times 10^{-16} g^{-1} cm^2 s^{-1}$.

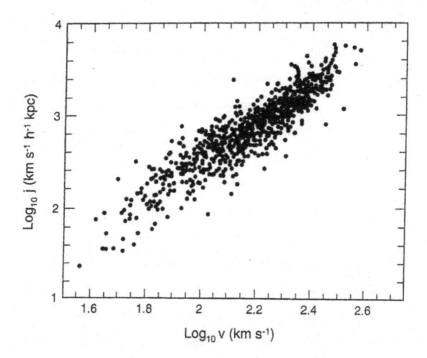

FIG. 4.2 – The specific angular momenta j ($\equiv J/M$) for the disks of spiral galaxies versus their rotation speed v. This plot is adapted from one by Steinmetz and Navarro (1999; h is a dimensionless measure of Hubble's parameter). The Tully-Fisher relation as revealed here needs modest assumptions to put it into correspondence with basic physics, as discussed by Bullock et al. (2001). But the data support $j = pM$ or $J = pM^2$, where p is a constant fixed by the larger class of data in Fig. 1.

of other kinds. There has been considerable discussion of its origin, since while it is compatible with standard gravitational theory it appears to require some additional factor to account for why it holds over such a large mass range. (It may or may not be a coincidence that the dimensionless combination G/pc is the same order of magnitude as the fine structure constant, though this supports the conjecture that p is the analog of h.) What we have shown here is that an angular momentum / mass relation of the form (4.35) might be expected on the basis of 5D theory.

4.6 Particles and Waves: A Rapprochement

It appears contradictory that something can behave as a particle in one state and a wave in another. The archetypal example is the double-slit experiment, where electrons as discrete particles pass through a pair of apertures and show wave-like interference patterns. Wave-particle duality is widely regarded as a conceptual conflict between quantum and classical mechanics. However, particles and waves can both be given geometrical descriptions, which raises the possibility that these behaviours are merely different representations of the same underlying geometry (i.e. isometries). We have in Chapter 2 seen that a coordinate transformation can change the appearance and application of a solution, and we have in Chapter 3 noted a solution where a flat 5D space can represent a wave in ordinary 3D space. In this section, we will consider flat manifolds of various dimensionalities, with a view to showing that a 4D de Broglie wave which describes energy and momentum is isometric to a flat 5D space.

2D manifolds, like the one which describes the surface of the Earth, are locally flat. A brief but instructive account of their iso-metries is given by Rindler (1977, p. 114; a manifold of any N is ap-proximately flat in a small enough region, and changes of coordinates that qualify as isometries should strictly speaking preserve the signa-ture). Consider, as an example, the line element $ds^2 = dt^2 - t^2 dx^2$. Then the coordinate transformation $t \to e^{i\omega t} / i\omega$, $x \to e^{ikx}$ causes the metric to read $ds^2 = e^{2i\omega t} dt^2 - e^{2i(\omega t + kx)} dx^2$, where ω is a frequency, k is a wave number and the phase velocity ω/k has been set to unity. It is clear from this toy example that a metric which describes a freely-moving particle (the proper distance is proportional to the time) is equivalent to one which describes a freely-propagating wave. For the particle, we can define its energy and momentum via $E \equiv m \, (dt / ds)$ and $p \equiv m(dx / ds)$. For the wave, $\tilde{E} \equiv m e^{i\omega t} (dt / ds)$ and $\tilde{p} \equiv m e^{i(\omega t + kx)} (dx / ds)$. In both cases, the mass m of a test particle has to be introduced *ad hoc*, a shortcoming which will be addressed below. The standard energy condition (4.17), in the form $m^2 = E^2 - p^2$, is recovered if the signature is $(+-)$. If on the other hand we have a Euclidean signature of the kind used in certain approaches to quantum gravity, it is instructive in the 2D case to consider the isometry $ds^2 = x^2 dt^2 + t^2 dx^2$. The transformation $t \to e^{i\omega t} / i\omega$, $x \to e^{ikx} / ik$ causes this to read $ds^2 = - (1/k)^2 \, e^{2i(\omega t + kx)} (dt^2 + dx^2)$, after the absorption of a phase velocity as above. Thus a particle metric becomes one with a conformal factor which resembles a wave function.

3D manifolds add little to what has been discussed above. It is well known that in this case the Ricci and Riemann-Christoffel tensors can be written as functions of each other, so the field equations bring us automatically to a flat manifold as before.

4D manifolds which are isotropic and homogeneous, but non-static, lead us to consider the FRW models. These have line elements given by

$$ds^2 = dt^2 - \frac{R^2(t)}{\left(1 + k\,r^2/4\right)^2}\left(dx^2 + dy^2 + dz^2\right) \quad , \quad (4.36)$$

where $R(t)$ is the scale factor and $k = \pm 1, 0$ defines the 3D curvature. (This should not be confused with the wave number.) In the ideal case where the density and pressure of matter are zero, a test particle moves away from a local origin with a proper distance proportional to the time. (I.e., $R = t$ above where the spatial coordinates *xyz* and $r \equiv \sqrt{x^2 + y^2 + z^2}$ are comoving and dimensionless.) This specifies the Milne model, which by the field equations requires $k = -1$. (We can think of this as a situation where the kinetic energy is balanced by the gravitational energy of a negatively-curved 3D space.) However, (4.36) with $R = t$ and $k = -1$ is isometric to Minkowski space (Rindler 1977, p. 205). Indeed, the Milne model is merely a convenient non-static representation of flat 4D space. In the local limit where $|r^2/4| \ll 1$, the *t*-behaviour of the 3D sections of (4.36) allows us to specify a wave via the same kind of coordinate transformation used in the 2D case. We eschew the details of this, since the same physics is contained in more satisfactory form if the dimensionality is extended.

5D manifolds which are canonical have simple dynamics and lend themselves to quantization, as we have seen. It is therefore natural that we should consider a 5D canonical analog of the 4D Milne model discussed in the preceding paragraph (Liko and Wesson 2005). We desire that it be Ricci-flat ($R_{AB} = 0$) and Riemann-flat ($R_{ABCD} = 0$). The appropriate solution is given by

$$dS^2 = \left(\frac{l}{L}\right)^2 dt^2 - \left[l\sinh\left(\frac{t}{L}\right)\right]^2 d\sigma^2 - dl^2 \quad . \quad (4.37)$$

Here the 3-space is the same as that above, namely $d\sigma^2 = (dx^2+dy^2+dz^2)(1 + kr^2/4)^{-2}$ with $k = -1$. That the time-dependence of the 3-space in (4.37) is different from that in (4.36) is attributable to their different dimensionalities. However, the local situation for (4.37) is close to that for (4.36). To see this, we note that for laboratory situations $t/L \ll 1$ in (4.37), so it reads

$$dS^2 \simeq \left(\frac{l}{L}\right)^2 dt^2 - \left(\frac{lt}{L}\right)^2 d\sigma^2 - dl^2 \quad . \quad (4.38)$$

In this, let us multiply by L^2 and divide by ds^2. Also, we take the null-path hypothesis, which for any canonical metric results in the constraint $(dl/ds) = \pm l/L$. Then (4.38) gives

$$0 \simeq l^2 \left(\frac{dt}{ds}\right)^2 - (lt)^2 \left[\left(\frac{dx}{ds}\right)^2 + \left(\frac{dy}{ds}\right)^2 + \left(\frac{dz}{ds}\right)^2\right] - l^2 \quad . \quad (4.39)$$

This with the identification $l = m$ (see before) and the recollection that proper distances are defined by $\int t dx$ etc., simply reproduces the standard condition (4.17) in the form $0 = E^2 - p^2 - m^2$.

To convert the 5D metric (4.38) to a wave, we follow the lower-dimensional examples noted before. Specifically, we change $t \rightarrow e^{i\omega t} / i\omega$, $x \rightarrow \exp(ik_x x)$ etc., where ω is a frequency and k_x etc. are wave numbers for the x, y, z directions. After setting the phase velocity to unity, (4.38) then reads

$$dS^2 \simeq \left(\frac{l}{L}\right)^2 e^{2i\omega t} dt^2 - \left(\frac{l}{L}\right)^2 \left\{\exp\left[2i(\omega t + k_x x)\right] dx^2 + \text{etc}\right\} - dl^2 \quad .$$

$$(4.40)$$

This with the null condition causes the analog of (4.39) to read

$$0 \simeq \left\{l e^{i\omega t} \frac{dt}{ds}\right\}^2 - \left\{l \exp\left[i(\omega t + k_x x)\right]\frac{dx}{ds}\right\}^2 - \text{etc} - l^2 \quad . \quad (4.41)$$

We can again make the identification $l = m$ and define

$$\tilde{E} \equiv l e^{i\omega t} \frac{dt}{ds}, \quad \tilde{p} \equiv l \exp\left[i(\omega t + k_x x)\right]\frac{dx}{ds} \quad \text{etc.} \quad (4.42)$$

Then (4.41) is equivalent to

$$0 \simeq \tilde{E}^2 - \tilde{p}^2 - m^2 \quad . \tag{4.43}$$

This is of course the wave analog of the standard relation (4.17) for a particle.

The fact that it is possible to convert a particle solution to a wave solution, with consistency of their energy relations, demonstrates in a formal sense that the concepts are compatible. It should also be pointed out that while we have done this only for (4.37), that solution is 5D flat. Many solutions have this property, so it must be possible to do the same thing for them, at least in principle. [In practice, it may be difficult to find the appropriate coordinate transformations. The wave of (3.22) found by Billyard and Wesson (1996) is also 5D flat, but has a different signature to (4.37), and a coordinate transformation between them is not known.] Furthermore, even 5D solutions which are globally curved are locally flat, so the correspondence outlined above has generality. We are led to conclude that as regards their description by 5D relativity, particles and waves can be regarded as isometries. That is, they are different 4D representations of the same flat 5D space.

4.7 Conclusion

Quantum and classical physics parted ways in the 1930s. Then, there were good experimental data on atomic systems, which could be adequately explained by the simple but effective theory provided by Schrodinger's wave equation and Heisenberg's account of quantum states. By contrast, cosmological observations of galaxies were sketchy, and Einstein's theory of general relativity was too complicated to be widely appreciated, let alone the extensions of it due to Kaluza and Klein. Unfortunately, knowledge has its own kind of inertia: the more one learns, the more one wishes to learn, usually along the same path. Therefore, more work was done on quantum mechanics than on general relativity. Interest in gravitation only accelerated in the 1960s, due largely to Wheeler, who pointed out that a proper understanding of condensed astrophysical objects could only be obtained via Einstein's theory. That progress was enlivened by Hoyle, who sprinkled astronomy with pregnant ideas; and Hawking, who made the idea of a black hole acceptable to a wide audience. However, it is the case that modern physics remains split into two camps: the quantum and the classical.

The contents of the current chapter have shown how this split might be mended by the device of a fifth dimension. We have in effect used an extra coordinate related to the rest mass of a particle to give a semi-classical account of several problems which are usually regarded as the purview of quantum theory. (Here "semi-classical" means a formalism which resembles Einstein's theory but contains Planck's constant.) The issues we have examined involve the following: The derivation of 4D uncertainty from the laws of a 5D deterministic world (Section 4.2); the possible existence of a mass quantum of small size, which is connected to the cosmological constant as a measure of the small curvature of the universe (Section 4.3); the demonstration that the Klein-Gordon and Dirac equations can be understood as consequences of 5D dynamics (Section 4.4); the realization that in 5D quantization is gauge-dependent, so a relation for a microscopic system like a particle might reappear as a relation for a macroscopic system like a galaxy (Section 4.5); and the possibility that a particle and a wave may be the same thing described in different coordinate frames, or isometries (Section 4.6). These issues do not, of course, exhaust the list of problems we might wish to solve. Rather, we have dealt with the issues which can be readily treated with one of the two natural gauges of the theory.

These gauges are those named after Einstein and Planck, where respectively the mass of a particle is measured by its Schwarzschild radius or its Compton wavelength. These gauges exist because of the historical development of physics, which in its bipolar concentration on gravitation and quantum mechanics has shown us the relevant constants involved (G and h). While these are only two out of an infinite number of coordinate frames allowed by the covariance of the theory, they are ideally suited to the physics in their respective domains. The situation here is similar to that in other covariant theories, like general relativity. In the latter, we recognize the Schwarzschild solution as relevant to the solar system when it is couched in its original coordinates, whereas other coordinates (like those due to Eddington-Finkelstein or Kruskal) may have analytical value but do not correspond directly to our observations. In this regard, it is apparent that the results derived in the present chapter have relevance mainly for induced-matter (or space-time-matter) theory, rather than membrane theory with its singular hypersurface. That said, in both versions of 5D relativity a central role is played by the cosmological "constant", or more correctly stated, by the energy density of the vacuum.

Indeed, both classical cosmology and quantum field theory now ascribe great importance to that part of the universe we cannot "see". It is to this which we now turn our attention.

References

Billyard, A., Wesson, P.S. 1996, Gen. Rel. Grav. 28, 129.

Bullock, J.S., Dekel, A., Kolatt, T.S., Kravtsov, A.V., Klypin, A.A., Porciani, C., Primack, J.R. 2001, Astrophys. J. 555, 240.

Desloge, E.A. 1984, Am. J. Phys. 52, 312.

Kaluza, T. 1921, Sitz. Preuss. Akad. Wiss. 33, 966.

Klein, O. 1926, Z. Phys. 37, 895.

Liko, T., Wesson, P.S. 2005, J. Math. Phys. 46, 062504.

Lineweaver, C.H. 1998, Astrophys. J. 505, L69.

Overduin, J.M. 1999, Astrophys. J. 517, L1.

Padmanabhan, T. 2003, Phys.Rep. 380, 235.

Perlmutter, S. 2003, Phys. Today 56 (4), 53.

Ponce de Leon, J. 2001, Mod. Phys. Lett. A16, 2291.

Rindler, W. 1977, Essential Relativity (2nd ed., Springer, Berlin).

Steinmetz, M., Navarro, J.F. 1999, Astrophys. J. 513, 555.

Weinberg, S. 1980, Rev. Mod. Phys. 52, 515.

Wesson, P.S. 1981, Phys. Rev. D23, 1730.

Wesson, P.S. 1999, Space-Time-Matter (World Scientific, Singapore).

Wesson, P.S. 2002, Class. Quant. Grav. 19, 2825.

Wesson, P.S. 2003, Gen. Rel. Grav. 35, 111.

Wesson, P.S. 2005, Astrophys. Sp. Sci., 299, 317.

Youm, D. 2000, Phys. Rev. D62, 084002.

Youm, D. 2001, Mod. Phys. Lett. A16, 2371.

5. THE COSMOLOGICAL "CONSTANT" AND VACUUM

"Sir, the invisible man is outside – but I said you couldn't see him"

(Hollywood gag)

5.1   Introduction

In all of the preceding chapters we have mentioned the cosmological "constant", which is a true constant in Einstein's 4D theory of general relativity, but a possibly variable measure of the properties of the vacuum in $N \geq$ 5D theories. In the present chapter, we wish to focus on Λ and the concept of vacuum in 5D.

The role of Λ as a scale for the universe was clearly perceived by Eddington, who was a sagacious student of general relativity at a time when Einstein's theory was not widely appreciated. Einstein's semi-technical book *The Meaning of Relativity* was based on the Stafford Little lectures at Princeton (New Jersey) in May 1921. Eddington's semi-popular volume *The Expanding Universe* was based on his International Astronomical Union lecture at Cambridge (Massachusetts) in September 1932 (though did not come out in accessible form til considerably later). The fact that both books are under 200 pages long, but have had profound impacts on physics, is a lesson that one does not need a tome of technicalities to convey the gist of a subject. It is also remarkable that these books and their authors held very different views about Λ. Einstein's distrust of this parameter is, of course, well known; whereas Eddington was led to state that "To drop the cosmical constant would knock the bottom out of space" (loc cit.,

p. 104). The latter worker, based largely on his studies in cosmology, regarded the relationship between the Einstein tensor and the metric tensor $G_{\alpha\beta} = \Lambda g_{\alpha\beta}$ as the basis of gravitation. He went on to use the fundamental length scale defined by Λ with Planck's constant, to argue that an uncertainty of the kind associated with Heisenberg meant that we can never know the precise momentum (and therefore location) of any particle in the universe. In this and other ways, Eddington presaged parts of what today we derive from the application of quantum field theory to curved spacetimes, which includes the Hawking radiation around black holes. He also predicted to order of magnitude the role played by the number of particles ($\approx 10^{80}$) in the visible part of the universe, an idea we now attribute mainly to Dirac who formalized it in the Large Numbers Hypothesis; and from this Eddington vaguely anticipated the constraints on evolution which nowadays we identify with Carter, Dicke and Hoyle in the form of the Anthropic Principle. All this from a belief in the cosmological constant and some very clear, succinct thinking.

We are endeavoring, in the present treatise, to emulate the Eddington approach. However, one aspect has come to the forefront since his time which plays a central part in modern thinking on Λ. Namely, that by the field equations, it measures the energy density and pressure of the vacuum via $\rho c^2 = -p = \Lambda c^4 / 8\pi G$ (in conventional units). That Λ as an explicit measure of scale coupled to the metric tensor, can be viewed instead as an implicit part of the energy-momentum tensor in the field equations $G_{\alpha\beta} = (8\pi G / c^4) T_{\alpha\beta}$, is due largely to the algebraic properties of the latter object. This was realized by Zeldovitch and others in the 1960s, and the usage is now commonplace via the relation just quoted. As elsewhere, however, an apparent conflict appears when the classical approach is extended to the quantum one. Modern quantum field theory involves vacuum fields, akin to the older zero-point fields, which can be measured by an effective value of Λ. Unfortunately, the values of Λ as inferred from astrophysics and particle physics differ. This is the crux of the cosmological-"constant" problem, which we do not wish to delve into here because good reviews are available (Weinberg 1980; Padmanabhan 2003). The offset is model-dependent, but in round figures is of the order of 10^{120}. This is a number which would have given even the numerically-minded Eddington pause for thought.

One of the most promising ways to account for the nature and various estimates of Λ is that it is a parameter which we measure in 4D but originates in $N \geq 5D$ (Rubakov and Shaposhnikov 1983; Csaki, Ehrlich and Grojean 2001; Seahra and Wesson 2001; Wesson and Liu 2001; Padmanabhan 2002; Mansouri 2002; Shiromizu, Koyama and Torii 2003; Mashhoon and Wesson 2004). There are different versions of this idea; but in its most general form it involves a reduction of field equations in $N \geq 5D$ to effective ones in 4D which contain a vacuum field which can be variable. For 4D models of cosmological type, like the FRW ones, this may only make the cosmological "constant" a time-variable parameter. (The physical dimensions of Λ in Einstein's equations are those of inverse-length squared, which means that it defines a distance of order 10^{28} cm; but via the speed of light this implies that it might be expected to decay as inverse-time squared, with a period of order 10^{10} yr.) In more complicated situations, it is possible in principle that Λ as the 4D measure of a 5D scalar field could vary in both time and space. This would resolve the cosmological-"constant" problem in a most satisfying manner.

In Section 5.2, we will look at a simple but instructive model where Λ is variable in a manner which is readily calculable (Mashhoon and Wesson 2004). This model is based on the generic property that physics which is covariant in 5D is gauge-dependent (via the extra coordinate) in 4D. We will use the induced-matter approach, but only as a mathematical framework, which implies applicability to other 5D formalisms (Ponce de Leon 2001). The model will use boundary conditions set by accepted wisdom concerning the big bang. But any approach which introduces effects due to the fifth dimension into standard 4D cosmology ought to predict more than it explains, so in Section 5.3 we will list consequences of the extra dimension for conventional astrophysics. These will be seen to provide no serious obstacle. So in Section 5.4 we will take up a more fundamental issue concerned with the gauge-dependence of Λ, namely the question of vacuum instability. We hasten to add that our model for this is of a modest kind, distinct from more radical ones which raise the possibility of a catastrophic destabilization of the vacuum due to high-energy experiments with particle accelerators. In 5D theory, there is no sharp division between what we call "vacuum" and what we call "matter", so a change in the former can lead to the creation of the latter, in accordance with the appropriate laws (Birrel and Davies 1982; Alvarez and Gavela 1983; Kolb, Lindley and Seckel 1984; Huang 1989; Linde 1991; Liko and Wesson 2005). The subject of vacuum instability is grave but speculative, so to balance things we return in Section 5.5 to the traditional subject of Mach's Principle, and use an exact solution of the 5D equations to show how it can be realized in 4D (Wesson, Seahra and Liu 2002). This presupposes, as do other considerations in this chapter, that we are willing to relax somewhat our conventional distinction between "matter" and "vacuum".

5.2   The 5D Cosmological "Constant"

In this section we absorb G, c and h, but the terminology is otherwise as before. Then the general 5D metric

$$dS^2 = g_{\alpha\beta}\left(x^\gamma, l\right) dx^\alpha dx^\beta + \varepsilon\Phi^2\left(x^\gamma, l\right) dl^2 \quad , \qquad (5.1)$$

describes gravity and a scalar field for both a spacelike and timelike extra dimension $\left(\varepsilon = \mp 1\right)$. As we saw in Chapter 1, for (5.1) the field equations $R_{AB} = 0$ can be expressed as sets of 10,4 and 1 relations, which we restate here for convenience:

$$G_{\alpha\beta} = 8\pi T_{\alpha\beta}$$

$$8\pi T_{\alpha\beta} \equiv \frac{\Phi_{\alpha;\beta}}{\Phi} - \frac{\varepsilon}{2\Phi^2} \left\{ \frac{\overset{**}{\Phi} \overset{*}{g}_{\alpha\beta}}{\Phi} - \overset{**}{g}_{\alpha\beta} + g^{\lambda\mu} \overset{*}{g}_{\alpha\lambda} \overset{*}{g}_{\beta\mu} \right.$$

$$\left. - \frac{g^{\mu\nu} \overset{*}{g}_{\mu\nu} \overset{*}{g}_{\alpha\beta}}{2} + \frac{\overset{*}{g}_{\alpha\beta}}{4} \left[g^{\mu\nu} \overset{*}{g}_{\mu\nu} \left(g^{\mu\nu} \overset{*}{g}_{\mu\nu} \right)^2 \right] \right\}$$ (5.2)

$$P^{\beta}_{\alpha;\beta} = 0$$

$$P^{\beta}_{\alpha} \equiv \frac{1}{2\Phi} \left(g^{\beta\sigma} \overset{*}{g}_{\sigma\alpha} - \delta^{\beta}_{\alpha} g^{\mu\nu} \overset{*}{g}_{\mu\nu} \right)$$ (5.3)

$$\varepsilon\Phi\Box\Phi = -\frac{g^{\lambda\beta} \overset{*}{g}_{\lambda\beta}}{4} - \frac{g^{\lambda\beta} \overset{**}{g}_{\lambda\beta}}{2} + \frac{\overset{*}{\Phi} g^{\lambda\beta} \overset{*}{g}_{\lambda\beta}}{2\Phi}$$

$$\Box\Phi \equiv g^{\alpha\beta}\Phi_{\alpha;\beta}$$ (5.4)

Here $\Phi_{\alpha} \equiv \partial\Phi/\partial x^{\alpha}$, a semicolon denotes the ordinary 4D covariant derivative and an overstar means $\partial/\partial x^4$. The canonical metric is obtained from (5.1) by factorizing it in terms of $x^4 = l$ and a constant length L:

$$dS^2 = \frac{l^2}{L^2} \left[g_{\alpha\beta}\left(x^{\gamma},l \right) dx^{\alpha} dx^{\beta} \right] - dl^2 \quad .$$ (5.5)

This form causes the Einstein tensor to read

$$G_{\alpha\beta} = \frac{1}{2L^2} \left[-l^2 \overset{**}{g}_{\alpha\beta} + l^2 g^{\lambda\mu} \overset{*}{g}_{\alpha\lambda} \overset{*}{g}_{\beta\mu} - 4l \overset{*}{g}_{\alpha\beta} - \frac{l^2}{2} g^{\mu\nu} \overset{*}{g}_{\mu\nu} \overset{*}{g}_{\alpha\beta} \right]$$

$$+\frac{1}{2L^2}\left[6+2l\,g^{\mu\nu}\,\overset{*}{g}_{\mu\nu}+\frac{l^2}{4}\,\overset{*}{g}{}^{\mu\nu}\,\overset{*}{g}_{\mu\nu}+\frac{l^2}{4}\left(g^{\mu\nu}\,\overset{*}{g}_{\mu\nu}\right)^2\right]g_{\alpha\beta} \quad . \quad (5.6)$$

In (5.5) the Weak Equivalence Principle is recovered as a symmetry via $\overset{*}{g}_{\alpha\beta}=0$ (Section 3.5). Then (5.6) gives $G_{\alpha\beta}=3g_{\alpha\beta}\,/\,L^2$, which are Einstein's equations with a cosmological constant $\Lambda=3\,/\,L^2$. We quoted this as (1.15). The analysis just given, and others of a similar type in the literature, show how the cosmological constant of Einstein theory is derived from Kaluza-Klein theory.

The preceding argument, however, begs for generalization. It is clear from (5.5) that this will involve a choice for $g_{\alpha\beta}=g_{\alpha\beta}\,(x^\gamma, l\,)$. Such a choice of gauge will not in general produce a constant Λ, so to this extent we expect a gauge-dependent cosmological "constant".

Let us look at a special but physically instructive case of (5.5). That metric is general, so to make progress we need to apply some physical filter to it. Now, the physics of the early universe is commonly regarded as related to inflation; and the standard 4D metric for this is that of de Sitter, where $ds^2=dt^2-\exp\left[2t\sqrt{\Lambda/3}\,\right]d\sigma^2$. (Here $d\sigma^2\equiv dr^2+r^2d\theta^2+r^2\sin^2\theta\,d\phi^2$ in spherical polar coordinates.) The physics flows essentially from the cosmological constant Λ. However, it is well known that the de Sitter metric is conformally flat. This suggests that physically-relevant results in 4D may follow from metric (5.5) in 5D if the latter is restricted to the 4D conformally-flat form:

$$dS^2 = \frac{l^2}{L^2}\left[f\left(x^\gamma, l\right)\eta_{\alpha\beta}dx^\alpha dx^\beta \right] - dl^2 \quad . \tag{5.7}$$

Here $\eta_{\alpha\beta}$ = diagonal $(+1 - 1 - 1 - 1)$ is the metric for flat Minkowski space. We are particularly interested in the l – dependence of $f(x^\gamma, l)$. To determine the latter, we need to solve the field equations.

We could take these in the form (5.2) – (5.4), but since we have suppressed the scalar field in (5.5) it is more convenient to calculate the components of the 5D Ricci tensor directly. These are:

$$R_{44} = -\frac{\partial A^\alpha_{\ \alpha}}{\partial l} - \frac{2}{l}A^\alpha_{\ \alpha} - A_{\alpha\beta}A^{\alpha\beta} \tag{5.8}$$

$$R_{\mu 4} = A^\alpha_{\ \mu\ ;\alpha} - \frac{\partial \Gamma^\alpha_{\ \mu\alpha}}{\partial l} \tag{5.9}$$

$${}^5R_{\mu\nu} = {}^4R_{\mu\nu} - S_{\mu\nu} \quad , \tag{5.10}$$

where $S_{\mu\nu}$ is a symmetric tensor given by

$$S_{\mu\nu} \equiv \frac{l^2}{L^2}\left[\frac{\partial A_{\mu\nu}}{\partial l} + \left(\frac{4}{l} + A^\alpha_{\ \alpha} \right)A_{\mu\nu} - 2A_\mu^{\ \alpha}A_{\nu\alpha} \right] + \frac{1}{L^2}\left(3 + lA^\alpha_{\ \alpha} \right)g_{\mu\nu} \quad . \tag{5.11}$$

Here ${}^4R_{\mu\nu}$ and $\Gamma^\mu_{\nu\rho}$ are, respectively, the 4D Ricci tensor and the connection coefficients constructed from $g_{\alpha\beta}$. Moreover

$$A_{\alpha\beta} \equiv \frac{1}{2}\frac{\partial g_{\alpha\beta}}{\partial l} \quad , \tag{5.12}$$

where $A_\alpha^\beta = g^{\beta\delta} A_{\alpha\delta}$. We need to solve (5.8) – (5.10) in the form $R_{AB} = 0$, subject to putting $g_{\mu\nu}(x^\gamma, l) = f(x^\gamma, l)\eta_{\mu\nu}$ as in (5.7), which ensures (4D) conformal flatness. We note that $g^{\mu\nu} = \eta^{\mu\nu}/f$ and $A_{\mu\nu} = \overset{*}{f}\eta_{\mu\nu}/2$, where $\overset{*}{f} \equiv \partial f(x^\gamma, l)/\partial l$. Also $A^{\alpha\beta} = \overset{*}{f}\eta^{\alpha\beta}/(2f^2)$, $A^\alpha_{\ \alpha} = 2\overset{*}{f}/f$ and $A^\alpha_{\ \beta} = \overset{*}{f}\eta^\alpha_{\ \beta}/(2f)$. Then the scalar component of the field equation (5.8) becomes

$$2\frac{\partial}{\partial l}\left(\frac{\overset{*}{f}}{f}\right) + \left(\frac{\overset{*}{f}}{f}\right)^2 + \frac{4}{l}\left(\frac{\overset{*}{f}}{f}\right) = 0 \quad . \tag{5.13}$$

To solve this, we define $U \equiv \overset{*}{f}/f + 2/l$. Then (6) is equivalent to $2\overset{*}{U} + U^2 = 0$, or $\partial (U^{-1})/\partial l = 1/2$, so on introducing an arbitrary function of integration $l_0 = l_0(x^\gamma)$ we obtain $U^{-1} = [l - l_0(x^\gamma)]/2$. This, in terms of the original function f, means that $\overset{*}{f}/f + 2/l = U = 2/[l - l_0(x^\gamma)]$, or $\partial[\ln(l^2 f)]/\partial l = \partial\{\ln[l - l_0(x^\gamma)]^2\}/\partial l$. This gives $l^2 f/[l - l_0(x^\gamma)]^2 = k(x^\gamma)$, where $k = k(x^\gamma)$ is another arbitrary function of integration. We have noted this working to illustrate that the solution of the scalar component of the field equations (5.8) or (5.13) involves an arbitrary length $l_0(x^\gamma)$ and an arbitrary dimensionless function $k(x^\gamma)$. The solution for the conformal factor in the metric $g_{\mu\nu}(x^\gamma, l) = f(x^\gamma, l)\eta_{\mu\nu}$ is

$$f\left(x^{\gamma},l\right) = \left[1 - \frac{l_0\left(x^{\gamma}\right)}{l}\right]^2 k\left(x^{\gamma}\right) \tag{5.14}$$

and involves both arbitrary functions.

However, one of these is actually constrained by the vector component of the field equations (5.9). To see this we note that $A_{\mu\nu}$ of (5.12) is symmetric, and it is a theorem that then

$$A^{\mu}_{\;\nu\,;\mu} = A^{\mu}_{\;\nu;\mu} = \frac{1}{\sqrt{-g}}\frac{\partial}{\partial x^{\mu}}\left(\sqrt{-g}\,A^{\mu}_{\;\nu}\right) - \frac{A^{\alpha\beta}}{2}\frac{\partial g_{\alpha\beta}}{\partial x^{\nu}} \quad . \tag{5.15}$$

Here g is the determinant of the 4D metric, so since $g_{\mu\nu} = f\eta_{\mu\nu}$ we have $\sqrt{-g} = f^2$. Then using (5.15), equation (5.9) becomes

$$\frac{1}{2f^2}\frac{\partial}{\partial x^{\mu}}\left(f\overset{*}{f}\delta^{\mu}_{\;\nu}\right) - \frac{\overset{*}{f}}{f^2}\frac{\partial f}{\partial x^{\nu}} = \frac{\partial}{\partial l}\left(\Gamma^{\alpha}_{\;\nu\alpha}\right) \quad . \tag{5.16}$$

The r.h.s. of this can be expressed using the identity $\Gamma^{\alpha}_{\;\nu\alpha} \equiv \left(\sqrt{-g}\right)^{-1}$ $\partial\left(\sqrt{-g}\right)/\partial x^{\nu}$, whence (5.16) becomes

$$\frac{1}{2f^2}\frac{\partial}{\partial x^{\nu}}\left(f\overset{*}{f}\right) - \frac{\overset{*}{f}}{f^2}\frac{\partial f}{\partial x^{\nu}} = 2\frac{\partial}{\partial l}\left[\frac{1}{f}\frac{\partial f}{\partial x^{\nu}}\right] \quad . \tag{5.17}$$

In this form, we can multiply by $2f^2$ and rearrange to obtain

$$f \frac{\partial \overset{*}{f}}{\partial x^{\nu}} = \overset{*}{f} \frac{\partial f}{\partial x^{\nu}} \quad . \tag{5.18}$$

Dividing by $f \overset{*}{f} \neq 0$, we find

$$\frac{\partial}{\partial x^{\nu}} \left(\frac{\overset{*}{f}}{f} \right) = 0 \quad . \tag{5.19}$$

But the term in parentheses here, by (5.14), is $\overset{*}{f} / f = 2 l_0 \left(x^{\gamma} \right) /$ $\left\{ l \left[l - l_0 \left(x^{\gamma} \right) \right] \right\}$. Thus (5.19) implies that $l_0 \left(x^{\gamma} \right) = l_0$ and is constant. We have noted this working to illustrate that the scalar and vector components of the field equations (5.8) and (5.9) together yield the conformal factor

$$f \left(x^{\gamma}, l \right) = \left(1 - \frac{l_0}{l} \right)^2 k \left(x^{\gamma} \right) \quad , \tag{5.20}$$

which involves only one arbitrary function that is easy to identify: if the constant parameter l_0 vanishes, then $k \eta_{\mu\nu}$ is simply our original de Sitter metric tensor.

The tensor component of the field equations (5.10) does *not* further constrain the function $k(x^{\gamma})$. However, we need to work through this component in order to isolate the 4D Ricci tensor ${}^4 R_{\mu\nu}$ and so obtain the effective cosmological "constant". To do this, we need to evaluate $S_{\mu\nu}$ of (5.11). The working for this is straightforward but tedious. The result is simple, however:

$$S_{\mu\nu} = \frac{3}{L^2} k\left(x^\gamma\right) \eta_{\mu\nu} \quad . \tag{5.21}$$

By the field equations (5.10) in the form $^5R_{\mu\nu} = 0$, this means that the 4D Ricci tensor is also equal to the r.h.s. of (5.21). We recall that our (4D) conformally-flat spaces (5.7) have $g_{\mu\nu} = f(x^\gamma, l) \; \eta_{\mu\nu} = (1 - l_0 / l)^2$ $k\,(x^\gamma)\eta_{\mu\nu}$ using (5.20). Thus $k\,(x^\gamma)\eta_{\mu\nu} = l^2 g_{\mu\nu} / (1 - l_0)^2$ and

$$^4R_{\mu\nu} = \frac{3}{L^2} \frac{l^2}{\left(1 - l_0\right)^2} g_{\mu\nu} \quad . \tag{5.22}$$

This is equivalent to the Einstein field equation for the de Sitter metric tensor $k\eta_{\mu\nu}$, since under a constant conformal scaling of a metric tensor, the corresponding Ricci tensor remains invariant. Nonetheless, (5.22) defines an Einstein space $^4R_{\mu\nu} = \Lambda g_{\mu\nu}$ with an effective cosmological constant given by

$$\Lambda = \frac{3}{L^2}\left(\frac{l}{1 - l_0}\right)^2 \quad . \tag{5.23}$$

This is our main result. It reduces for $l_0 = 0$ to the standard de Sitter value $\Lambda = 3 / L^2$. The latter, as we showed above, holds when there is no l-dependence of the 4D part of the canonical metric (5.5). By contrast, when there is an l-dependence of the form given by (5.20) we obtain (5.23). The difference between the Λ forms is mathematically modest, but can be physically profound, because (5.23) says that for $l \to l_0$ we have $\Lambda \to \infty$. In other words, the cosmological constant is not only gauge-dependent but also divergent for a certain value of the extra coordinate.

This is a striking result, and at first sight puzzling. However, we should recall that if we have a theory which is covariant in 5D and from it derive a quantity which is 4D in nature, then in general a change in the 5D coordinate frame will alter the form of the 4D quantity. In our case, the field equations $R_{AB} = 0$ are clearly covariant, and we have changed the coordinate frame away from the canonical one and found that we have altered the form of Λ. We can sum up the situation as follows. The pure-canonical metric and the conformally-flat metric have line elements given respectively by

$$dS^2 = \frac{l^2}{L^2} g_{\alpha\beta}\left(x^\gamma\right) dx^\alpha dx^\beta - dl^2 \qquad (5.24)$$

$$dS^2 = \frac{(l-l_0)^2}{L^2} k\left(x^\gamma\right) \eta_{\alpha\beta} dx^\alpha dx^\beta - dl^2 \qquad . \qquad (5.25)$$

Clearly the two are compatible, and the first implies the second if we shift $l \rightarrow (l - l_0)$ and write $g_{\alpha\beta}(x^\gamma) = k(x^\gamma)\eta_{\alpha\beta}$. The shift along the l-axis does not alter the last part of the metric, so both forms describe flat scalar fields. But the shift alters the prefactor on the first part of the metric, with the consequence that Λ changes from $3 / L^2$ to $(3 / L^2)$ $l^2 (l - l_0)^{-2}$ as in (5.23) above.

To investigate this in more detail, we will adopt the strategy of Chapter 3. There we saw that there is often an extra force per unit mass or acceleration which acts in 4D when a path is geodesic in 5D (the pure canonical metric is an exception). Also, we saw that the 5D path may be null. To proceed, we return to the general form of the metric (5.5), for which the path splits naturally into a 4D part and an extra part:

$$\frac{d^2 x^\mu}{ds^2} + \Gamma^\mu_{\alpha\beta} \frac{dx^\alpha}{ds} \frac{dx^\beta}{ds} = f^\mu$$

$$f^\mu \equiv \left(-g^{\mu\alpha} + \frac{1}{2} \frac{dx^\mu}{ds} \frac{dx^\alpha}{ds} \right) \frac{dl}{ds} \frac{dx^\beta}{ds} \frac{\partial g_{\alpha\beta}}{\partial l} \tag{5.26}$$

$$\frac{d^2 l}{ds^2} - \frac{2}{l} \left(\frac{dl}{ds} \right)^2 + \frac{l}{L^2} = -\frac{1}{2} \left[\frac{l^2}{L^2} - \left(\frac{dl}{ds} \right)^2 \right] \frac{dx^\alpha}{ds} \frac{dx^\beta}{ds} \frac{\partial g_{\alpha\beta}}{\partial l} \quad . \tag{5.27}$$

In these, following (5.25) and the preceding discussion of metrics, we substitute

$$g_{\alpha\beta}\left(x^\gamma, l\right) = \left(\frac{l - l_0}{l}\right)^2 k_{\alpha\beta}\left(x^\gamma\right) \quad , \tag{5.28}$$

where $k_{\alpha\beta}$ is any admissible 4D vacuum metric of general relativity with a cosmological constant $3 / L^2$. Furthermore we assume a null 5D path as noted above, and rewrite the line element as

$$dS^2 = \left[\frac{l^2}{L^2} - \left(\frac{dl}{ds} \right)^2 \right] ds^2 = 0 \quad . \tag{5.29}$$

Since a massive particle in spacetime has $ds^2 \neq 0$, we have that the velocity in the extra dimension is given by $(dl / ds)^2 = (l / L)^2$. Then the r.h.s. of (5.27) disappears, and to obtain the *l*-motion we need to solve

$$\frac{d^2l}{ds^2} - \frac{2}{l}\left(\frac{dl}{ds}\right)^2 + \frac{l}{L^2} = 0 \qquad (5.30)$$

and $(dl / ds)^2 = (l / L)^2$ simultaneously. Substituting the latter into (5.30), we find that *l* is a superposition of simple hyperbolic functions. There will be two arbitrary constants of integration involved in this solution, which can be written as

$$l = A\cosh\left(\frac{s}{L}\right) + B\sinh\left(\frac{s}{L}\right) \qquad . \qquad (5.31)$$

Moreover, $(dl / ds)^2 = (l / L)^2$ implies that $A^2 = B^2$. To fix the constants *A* and *B* here, it is necessary to make a choice of boundary conditions. It seems most natural to us to locate the big bang at the zero point of proper time and to choose $l = l_0$ ($s = 0$). Then $A = l_0$ and $B = \pm l_0$ in (5.31), which thus reads

$$l = l_0 e^{\pm s/L} \qquad . \qquad (5.32)$$

The sign choice here is trivial from the mathematical perspective, and merely reflects the fact that the motion is reversible. However, it is not trivial from the physical perspective, because it changes the behaviour of the cosmological constant.

This is given by (5.23), which with (5.32) yields

$$\Lambda = \frac{3}{L^2} \frac{1}{\left(1 - e^{\mp s/L}\right)^2} \qquad . \qquad (5.33)$$

In the first case (upper sign), Λ decays from an unbounded value at the big bang ($s = 0$) to its asymptotic value of $3 / L^2$ ($s \to \infty$). In the second case (lower sign), Λ decays from an unbounded value ($s = 0$) and approaches zero ($s \to \infty$). We infer from astrophysical data that the first case is the one that corresponds to our universe.

To investigate the physics further, let us now leave the last component of the 5D geodesic (5.27) and consider its spacetime part (5.26). We are especially interested in evaluating the anomalous force per unit mass f^μ of that equation, using our metric tensor (5.28). The latter gives $\partial g_{\alpha\beta} / \partial l = 2(l - l_0)(l_0 / l^3) k_{\alpha\beta} (x^\gamma) = 2l_0 [l (l - l_0)]^{-1} g_{\alpha\beta}$ in terms of itself. We can substitute this into (5.26), and note that the 4-velocities are normalized as usual via $g_{\alpha\beta} (dx^\alpha / ds)(dx^\beta / ds) = 1$. The result is

$$f^\mu = -\frac{l_0}{l(l - l_0)} \frac{dl}{ds} \frac{dx^\mu}{ds} \qquad . \qquad (5.34)$$

This is a remarkable result. It describes an acceleration in spacetime which depends on the 4-velocity of the particle and whose magnitude (with the choice of boundary conditions noted above) is infinite at the big bang. It is typical of the non-geodesic motion found in other applications of induced-matter and membrane theory. It follows from (5.32) that

$$f^\mu = \mp \frac{1}{L} \frac{dx^\mu}{ds} \frac{1}{\left(e^{\pm s/L} - 1\right)} \quad . \quad (5.35)$$

In the first case (upper sign), $f^\mu \to (-1 / s) (dx^\mu / ds)$ for $s \to 0$ and $f^\mu \to 0$ for $s \to \infty$. In the second case (lower sign), $f^\mu \to (-1 / s) (dx^\mu / ds)$ for $s \to 0$ and $f^\mu \to (-1 / L) (dx^\mu / ds)$ for $s \to \infty$. Thus both cases have a divergent, attractive nature near the big bang. However, at late times the acceleration disappears in the first case, but persists (though is small if L is large) in the second case. As in our preceding discussion of Λ, we infer from astrophysical data that the first case is the one that corresponds to our universe.

In (5.33) for $\Lambda = \Lambda(s)$ and (5.35) for $f^\mu = f^\mu(s)$ we have relations which form an interface between theory and observation. Indeed, we have already chosen the signs in our relations by appeal to the broad aspects of data on cosmological timescales and the dynamics of galaxies (Overduin 1999; Strauss and Willick 1995). However, there are more detailed comparisons and predictions which can be made.

5.3   Astrophysical Consequences

The model derived in the preceding section has a time-dependent cosmological "constant" $\Lambda = (3 / L^2) (1 - e^{-s/L})^{-2}$ given by (5.33) and a velocity-dependent fifth force $f = - (vc/L) (e^{s/L} - 1)^{-1}$ given by (5.35). Here s is the 4D proper time, v is the 3D velocity which we take to be radial (so $v \ll c$ implies $s \ll c\,t$), and L is a length we take to be 1×10^{28} cm approximately (see above). We reintroduce conventional units for the speed of light c and the gravitational constant G, because we wish to make some comments about the physics of the model with a view to comparison with observational data (Wesson 2005). In this regard, it should be recalled that all of the standard FRW models can be written in 4D conformally-flat form (Hoyle and Narlikar 1974), so the following 6 comments are expected to have some generality.

(a) The 5D model is like 4D inflationary ones, insofar as it is dominated by Λ at early times. Indeed, by the specified relation, Λ formally diverges near the big bang. To illustrate the potency of this, we note that over the period 10^8 to 10^{10} yr the value of Λ decreases by a factor of approximately 4000. From the viewpoint of general relativity, Λ has associated with it a force (per unit mass) $\Lambda rc^2 / 3$ and an energy density $\Lambda c^4 / 8\pi G$. It may be possible to test for the decay of these using high-redshift sources such as QSOs.

(b) Galaxy formation is augmented in this model, because there is a velocity-dependent extra force which tends to pull matter back towards a local origin. In the usual 4D scheme, galaxy formation is presumed to occur when an over-dense region attracts more material than its surroundings, so that the density perturbation grows with time. Unfortunately, it does not do so fast enough in most models to account for the observed galaxies or other structures (Padmanabhan 1993). And while there are ways out of this dilemma (for example by using seed perturbations due to quantum effects or pregalactic stars), the fifth force naturally aids galaxy formation and deserves in-depth study.

(c) Peculiar velocities are naturally damped by the fifth force noted above. It is standard in modern cosmology to break the velocities of galaxies into two components, the (regular) Hubble flow and (random) departures from it. In theoretical work, the former component is often removed by a choice of coordinates, defining a comoving frame where the regular velocities of the galaxies are zero and to which their peculiar velocities can be referred. A practical definition of the comoving frame is the one in which the 3K microwave background looks completely homogeneous. At the present epoch, the peculiar velocities of field galaxies do not exceed a few 100 km/s; and while there are other ways to account for this, the fifth force provides a natural mechanism. This can be appreciated by noting that the velocity associated with the force is $v = v_0(e^{s/L} - 1)^{-1}$, where v_0 is the value when $s = L \ln(2)$. This is asymptotic to zero, and is like the damped motion characteristic of a toy model based on 5D Minkowski space (Wesson 1999 pp. 169 – 172). In both cases, the motion has a form which is due essentially to our use of the 4D proper time $s \simeq ct$ as parameter. We discussed the consequences of this in Chapter 3, and in the present context we can express the motion in an alternative way: the comoving frame which is assumed in most 4D work on cosmology is a natural equilibrium state of 5D gravity.

(d) The damping mechanism outlined above can lead to a cosmological energy field which is significant. We can calculate an approximate upper limit to this by using previous expressions in the following manner. The magnitude of the force on an object of mass m

is $(mvc / L)(e^{s/L} - 1)^{-1}$. The product of this with v gives the associated rate of change of energy or power, which is $(mcv_0^2 / L) (e^{s/L} - 1)^{-3}$. The integral of this over proper time from s_1 to s_2 gives the total energy change, and if we assume $s_1 \ll s_2$ and $s_1 \ll L$ this is approximately $(mv_0^2 / 2) (L / s_1)^2$. This is the energy lost by one object through the damping of its kinetic energy by the fifth force, and is sharply peaked at early epochs. If the objects concerned form a uniform distribution, and presently have a mean distance d from each other, the energy density of the field produced by the damping is $\varepsilon \simeq \left(mv_0^2 / 2d^3\right)\left(L / s_1\right)^2$. We can write this in a more instructive form if we introduce the mass density (ρ_m), the epoch when the damping was severe (t_*) and the epoch now (t_0). Then a rough estimate of the present energy density of the field that results from the damping is $\varepsilon \simeq \left(\rho_m v_0^2 / 2\right)\left(t_0 / t_*\right)^2$. This is a theoretical upper limit because of the approximations made, but of course the parameters in it are themselves highly uncertain. For the purpose of illustration, let us substitute $\rho_m = 2 \times 10^{-31}$ g cm^{-3}, $v_0 = 100$ km/s, $t_0 = 1 \times 10^{10}$ yr and $t_* = 1 \times 10^8$ yr. Then $\varepsilon = 1 \times 10^{-13}$ erg cm^{-3}. This is a significant fraction of the energy density of the cosmic microwave background, and comparable to the energy densities of other components of the intergalactic photon field (Henry 1991). However, the physics which leads to the present field is different from that involved in the production of the CMB and the electromagnetic fields at other wavelengths such as the extragalactic background light (Wesson 1991). Indeed, the energy field currently being discussed may not be electromagnetic in nature (this would require that the damping act on plasma protogalaxies or young galaxies with a significant content of ionized material).   It could have a different nature, such as thermal energy or gravitational waves. We should recall, though, that while the nature of the energy field is open to discussion, its existence follows necessarily from the model being discussed. In view of this, we suggest that it would be wise to use observations of known fields and their isotropy to con-strain the underlying theory.

(e) The extra force associated with 5D strengthens local grav-ity and can therefore have dynamical effects on field galaxies and galaxies in clusters. This is because in the local limit Newton's law is modified so that the force (per unit mass) is

$$F = -\frac{GM}{r^2} - \frac{vc}{L} \frac{1}{\left(e^{ct/L} - 1\right)} \quad . \tag{5.36}$$

Here $M = M(r)$ is the mass interior to radius r for a system with ap-proximately spherical symmetry and we have put $s = ct$ in the fifth-force part. In view of the many implications of (5.36), we would like to discuss them in a form which is generic. One way to do this is to rewrite (5.36) as if the system was Newtonian and redefine the gravi-tational "constant" to be

$$G' = G\left[1 + \frac{vr^2}{GM}\left(\frac{c}{L}\right)\frac{1}{\left(e^{ct/L} - 1\right)}\right] \quad . \tag{5.37}$$

In this, vr^2 / GM has the dimensions of a time and would conventionally define a dynamical timescale (t_d), while L / c is a timescale associated with the cosmology which we expect to be approximately equal to the present epoch (t_0). Both this and the remaining factor in (5.37) are uncertain, so to be general we write the latter relation as

$$G' = G\left[1 + \alpha\left(\frac{t_d}{t_0}\right)\right] \quad . \qquad (5.38)$$

Here α is a dimensionless factor which depends on parameters to do with both the cosmology and the system, but which at present is of order of magnitude unity. In adopting this approach, it should be emphasized that we are not suggesting that the Newtonian "constant" is really a variable parameter, as in 4D gravitational theories of the type proposed by Brans / Dicke, Dirac, Hoyle / Narlikar, Canuto and others, or in certain $N (\geq 5)$D theories of the Kaluza / Klein type. Rather, we are taking a pragmatic approach to see how a new effect fits into a framework of existing data. In this respect, the parametization (5.38) also has the advantage that we can use limits set on departures from Newtonian gravity in other contexts (Will 1993). Since the relations (5.36) – (5.38) will require detailed future study, we content ourselves here with noting the results for two situations of interest. First, field galaxies which interact via their quadrupole moments at early times do so with an effective value of G which is 4 times the conventional one, thus largely resolving the discrepancy found in standard applications of this mechanism for the generation of the spin angular momenta of spirals (Hoyle 1949; Wesson 1982). Second, galaxies in clusters which interact over periods comparable to the crossing time do so with an effective value of G which is only modestly larger than the conventional one, thus only fractionally resolving the virial discrepancy found in many clusters and implying that most harbour large amounts of dark matter / energy, as usually assumed.

(f) The observed solar system is believed to be dynamically in agreement with 4D gravitational theory, with the possible exception of one situation to which we will return below. Before proceeding to this, it is instructive to recall that while Campbell's theorem guarantees the embedding of any 4D Einstein solution in a Ricci-flat 5D solution (Section 1.5), it does *not* guarantee that the 4D Birkhoff theorem carries over to 5D. Since Birkhoff's theorem, which ensures the uniqueness of the Schwarzschild solution (up to coordinate transformations) depends not only on the assumption of (3D) spherical symmetry but also on boundary conditions at infinity, it is perhaps not surprising that it breaks down when the theory is extended to 5D. This is why there are (at least) two solutions of the 5D theory, both of which agree with the observed dynamics of the solar system. One of these, called the 5D canonical Schwarzschild solution, has exactly the same dynamics as the 4D solution (Wesson 1999 pp. 177 – 179). The other, called the 5D soliton solution, has unmeasurably small departures from the 4D solution (Kalligas, Wesson and Everitt 1995). Both of these solutions are spherically symmetric in the three dimensions of ordinary space, and also static. By comparison, the cosmological model we are considering is non-static, principally by virtue of the time-dependent cosmological "constant" (5.33). This leads to the extra force (5.35), which affects the *radial* motion and depends critically on the ordinary velocity v in that direction. Clearly, if we have any prospect of seeing a cosmological 5D effect in the solar system, we have to look towards a situation where a test body has a large radial velocity. The planets, in their slow elliptical orbits, do not satisfy this criterion, and are expected to show no significant departures from standard motion. Other objects in the solar system, such as the particles of the solar wind or comets on parabolic orbits, do meet the criterion but are not well studied. By contrast, the Pioneer spacecraft are suitable (Anderson et al. 1998, 2002). These two craft, launched more than thirty years ago, have approximately radial orbits: Pioneer 10 is on a path just 3° out of the ecliptic, while Pioneer 11 is moving out of the ecliptic at about 17° inclination. At a distance of over 20 AU from the Sun, they are indicating an anomalous acceleration of negative sign of about 10^{-7} cm s^{-2}. Many possible explanations of this have been discussed, of which several are instrument-related (see Bertolami and Paramos 2004 for a review). Among those which are astrophysics-related, a plausible one involves an acceleration of the Sun due to its own asymmetric activity, but this falls short of explaining the anomaly by 4 orders of magnitude (Bini, Cherubini and Mashhoon 2004). By coincidence, the force (5.35) fails to account for the motion of these spacecraft by approximately the same factor. (The escape velocity from the solar system at 20 AU is close to 10 km

s^{-1}, and with this value for v and $L \simeq 1 \times 10^{28}$ cm, the result is as noted to within order of magnitude.) Nevertheless, we see here the opportunity for a future test of the 5D extra force. We need a high-velocity, radially-moving spacecraft. Better still would be two such craft, from which other influences could be cancelled as in the GRACE project (Tapley et al. 2004), leaving the cosmological effect we wish to verify.

The preceding comments (a) – (f) show that the model of Section 5.2, with a decaying cosmological "constant" (5.33) and an extra force (5.35), has wide-ranging consequences for astrophysics and cosmology. Indeed, the options are almost intimidating in their number, like the alternatives to the big bang we studied in Chapter 2. However, unlike some other forays into extra dimensions, the basic extension to 5D has the redeeming feature of testability.

5.4   Vacuum Instability

This possibility is raised by the considerations of the two preceding sections, where we saw that a translation along the extra axis of a 5D manifold can cause a gauge change in the effective value of the cosmological "constant" in 4D spacetime. This raises the question of changes in the energy density of the vacuum, which in general relativity is $\Lambda c^4 / 8\pi G$. We will give an alternative derivation of the expression for a variable Λ, and then outline the implications of this. The subject is speculative, so our discussion will be brief.

To see that Λ can depend on the extra coordinate of 5D theory ($x^4 = l$), let us reconsider the canonical form of the line element.

This has $dS^2 = (l/L)^2 ds^2 - dl^2$, where L is a length and $ds^2 = g_{\alpha\beta}(x^\gamma) dx^\alpha dx^\beta$ specifies the 4D proper time. The coordinate transformation or gauge change $l \to (l - l_0)$ leaves the extra part of the 5D metric unchanged, while the prefactor on the 4D part changes from l^2/L^2 to $(l - l_0)^2/L^2 = (l^2/L^2)[(l - l_0)/l]^2$. This means in effect that the original metric tensor $g_{\alpha\beta}$ changes to $\overline{g}_{\alpha\beta} = \left[(l-l_0)/l\right]^2 g_{\alpha\beta}$. Now it is a theorem that solutions of the source-free 5D field equations $R_{AB} = 0$ with the canonical metric satisfy the source-free 4D field equations ${}^4R_{\alpha\beta} = \Lambda g_{\alpha\beta}$ with $\Lambda = 3/L^2$ (see above and Mashhoon, Liu and Wesson 1994). It is also true that a constant conformal transformation of the metric leaves the Ricci tensor invariant, which is in effect the situation here since the change from $g_{\alpha\beta}$ to $\overline{g}_{\alpha\beta}$ only depends on l and not x^γ. Then the 4D field equations still hold with $\overline{R}_{\alpha\beta}(\overline{g}_{\alpha\beta}) = \overline{\Lambda}\,\overline{g}_{\alpha\beta}$ and $\overline{\Lambda} = (3/L^2)l^2/(l-l_0)^2$. This is the same as (5.23) above.

We see that a translation along the l-axis preserves the form of the canonical metric, and since the 5D field equations are covariant we obtain again the 4D field equations, but with a different cosmological constant, namely

$$\Lambda = \frac{3}{L^2}\left(\frac{l}{l-l_0}\right)^2 . \tag{5.39}$$

We examined the astrophysical consequences of this in Section 5.3, where we identified the divergence at $l = l_0$ with the big bang. We now proceed to take another look at (5.39), by adding a series of mathematical and physical conditions to it.

Firstly, let us take derivatives of (5.39) to obtain $d\Lambda = -(6/L^2)(l-l_0)^{-3}ll_0 dl$. We are mainly interested in the region near $l = l_0$, where the energy density $\Lambda = \Lambda(l)$ is changing rapidly but smoothly. Putting $dl = l - l_0$ for the change in the extra coordinate, we obtain

$$d\Lambda dl^2 = -6l^2/L^2 \quad . \tag{5.40}$$

This is an alternative form of the instability inherent in (5.39) near to its divergence.

Secondly, let us assume that the instability has a dynamical origin, and that the l–path involved is part of a null 5D geodesic as before. Then with $dS^2 = 0$ we have $l = l_0 e^{\pm s/L}$ as in (5.32). This with the upper sign implies $dl/l = ds/L$, which in (5.40) yields

$$d\Lambda ds^2 = -6 \quad . \tag{5.41}$$

This is remarkable, in that it contains no reference to $x^4 = l$ and is homogeneous in its physical dimensions (units), with no reference to fundamental constants. That is, it is gauge and scale invariant. [An alternative derivation of (5.41) may be made by using (5.33) and noting that s is measured from where Λ diverges at the big bang.] Again (5.41) implies instability, since $d\Lambda \to \infty$ for $ds \to 0$. This behaviour can be put into better physical perspective by recalling that the action for a particle of rest mass m in 4D dynamics is usually defined as

$I = \int mcds$. So (5.41) can be interpreted as a change in the energy density of the vacuum for a particle of unit mass which changes its action.

Thirdly, let us assume that the action is quantized. In some higher-dimensional theories, the rest mass of a particle can change as it pursues its 4D path, so $m = m\,(s)$. But irrespective of this, we have that $dI = mcds = h$, introducing Planck's constant. Then (5.41) gives

$$d\Lambda = -6\left(\frac{mc}{h}\right)^2 \quad . \tag{5.42}$$

This says that a change in the energy density of the vacuum (Λc^4 / $8\pi G$) is related to the square of the mass of a particle (m). It is a rather strange relation, in that the l.h.s. is classical in nature and the r.h.s. is quantum in nature. However, relations of a similar type have appeared in the literature (Matute 1997; Liu and Wesson 1998; Seahra and Wesson 2001; Mansouri 2002). We are reminded of the old Dirac theory, in which an underlying sea of energy develops a hole which is interpreted as a positron. On this basis, (5.42) can be interpreted to mean that a perturbation in the energy density of a global sea of energy has a size related to the Compton wavelength of an associated particle whose mass is m. Of course, in modern particle physics the masses are usually thought of as arising from a mechanism involving the Higgs field. We do not venture into this or related issues, because (5.42) is derived from classical as opposed to quantum theory.

In fact, the relations (5.40), (5.41) and (5.42) are phenomenological, in that they are semi-classical and lack a deeper foundation in quantum mechanics. They are akin to the relations of thermodynamics, which are compatible with – but lack detailed knowledge of – atomic physics. Nevertheless, (5.41) in particular is remarkably simple and deserves study.

5.5   Mach's Principle Anew

Most works on gravitation mention this subject, if only as a motivation for Einstein's general relativity. The latter is an excellent theory of gravity, but lacks a proper account of the source of that field, namely mass. It is possible to approach this subject from many different directions, which it would be inappropriate to review here. But to most workers, Mach's Principle means that the local properties of a massive particle are dependent on the nature and distribution of other matter in the universe. This has a nice, philosophical ring to it. However, it is indisputably the fact that the environment in which a particle finds itself is the vacuum. Therefore, it is reasonable to expect that any future theory of (say) the masses of the elementary particles will involve the physics of the vacuum. In the preceding section, we came across some unusual relations which appear to link the properties of the global vacuum to the mass of a local particle. In the present section, we wish to give a brief account of a similar subject from a different direction. Specifically, we wish to go back to a known wave-like solution in 5D (Liu and Wesson 1998), and take its

4D part (Wesson, Liu and Seahra 2002). This will be seen to have Machian properties.

We use the same approach as before, where a particle with energy E, ordinary momentum p and rest mass m may alternatively be regarded as a wave spread through spacetime, with the corresponding de Broglie and Compton wavelengths (see Sections 3.2 and 4.6). We retain conventional units as in Section 5.4, and as there we are interested in connecting the particle labels (E, p, m) to the properties of spacetime. As the basic descriptors for the latter, we take the components of the (4D) Ricci tensor R_β^α. From this we can obtain the Einstein tensor G_β^α, and from this we can construct if we wish an energy-momentum tensor T_β^α in accordance with Einstein's field equations.

The procedure as outlined so far runs parallel to that used for the induced-matter approach to fluids (see Chapter 1); except that now we are attempting to do it purely in 4D, and for a particle. Because of these constraints, we run into two technical issues: the metric turns out to be complex, and the energy-momentum tensor has an unusual form. Some workers may consider these issues to be problems. However, the subject we are addressing is essentially a technical one, namely: how to go from a particle, to a wave, to a spacetime. The fact that there is a wave in the middle of the analysis should alert us to the possibility of a complex metric; and the fact that we wish to obtain a particle (rather than fluid) description should alert us to the possibility that quantum (rather than classical) parameters may appear in the effective energy-momentum tensor. Due to these unorthodox properties, we keep the analysis short.

Consider, therefore, the simple case of a "wavicle" moving along the z-axis. The 4D line element is given by

$$ds^2 = c^2 dt^2 - \exp\left[2i\left(Et - pz\right)/h\right]dx^2$$

$$- \exp\left[-2i\left(Et - pz\right)/h\right]dy^2 - dz^2 \quad . \tag{5.43}$$

Despite the complex nature of the metric coefficients here, it transpires that the corresponding components of the Ricci tensor are real, as for certain other wave-like solutions (Section 4.6). The non-vanishing components for (5.43) are:

$$R_0^0 = \frac{2E^2}{h^2 c^2}, \quad R_3^0 = -\frac{2Ep}{h^2 c}, \quad R_3^3 = -\frac{2p^2}{h^2} \quad . \tag{5.44}$$

The trace of these is $R = 2\left(E^2 - p^2 c^2\right)/h^2 c^2$. Here the energy and momentum are constants, so we can introduce the mass in accordance with the standard relation $E^2 - p^2 c^2 = m^2 c^4$ of (3.5). Then we obtain $R = 2\left(mc/h\right)^2$. We see that the Ricci scalar is related to the Compton wavelength of the particle (h/mc), while the components of the Ricci tensor are related to its de Broglie wavelengths (hc/E, h/p). If we use the components (5.44) to construct G_β^α and so obtain T_β^α, the latter depends on particle parameters and Planck's constant, rather than fluid parameters and Newton's constant as usual (see Wesson, Liu and Seahra 2002). In a formal sense, (5.43) and (5.44) show how a local particle can be viewed as a global wave.

The solution (5.43) is special, in that it describes a wave which moves only along the z-axis and has E, p constant so that m is also constant. However, it is not difficult to see how to generalize the approach while keeping its main features. Thus a general correspondence between the geometry and the particle properties would be given by a relation of the form $R_{\alpha\beta} = -6\varepsilon^2 p_\alpha p_\beta / h^2$, where ε is a dimensionless coupling constant (Wesson 1999 p. 197). We have examined the dynamics which follow from such a relation. The usual law is modified to read $p_\alpha p^\alpha{}_{;\beta} = m(\partial m / \partial x^\beta)$, which gives back the 4D geodesic if the mass is constant. The 4-momenta are conserved via $p^\beta{}_{;\beta} = 0$, even if $m = m(x^\alpha)$. The last equation, if it could be evaluated, would effectively realize Mach's principle by telling us how to calculate the mass of a particle as a function of the coordinates. What we have seen here is that this principle is indeed compatible with general relativity, provided we are willing to stretch our understanding of metric and matter.

5.6   Conclusion

The cosmological "constant", when viewed from 5D, takes on a drastically different nature from the true constant of 4D. The canonical coordinates introduced by Mashhoon treat $x^4 = l$ in a way analogous to $x^0 = ct$ in the Robertson-Walker metric. When the 4D subspace of the 5D manifold does not depend on l, we recover the Weak Equivalence Principle as a geometric symmetry and the standard result $\Lambda = 3 / L^2$. Here the cosmological constant is a true constant, which measures the curvature of the embedded 4-space in the manner envisioned by Eddington (Section 5.2). However, if we carry out the simplest of coordinate transformations wherein $l \to (l - l_0)$, in the context of the simplest cosmological models which are like those of de Sitter, then we find that Λ changes by a factor $l^2 (l - l_0)^{-2}$ as in (5.23). This corresponds for a null 5D path to a decaying Λ, as in (5.33). The fifth force, which is generic to 5D relativity, then takes the form (5.35). There is a velocity-dependent acceleration which affects all objects in the universe, but which decreases in tandem with Λ as cosmic time increases. This has numerous astrophysical consequences, of which a half-dozen are accessible to observation (Section 5.3). The effect on Λ which follows from a translation in l has more general implications (Section 5.4). Most notably, there is a relation between the change in the cosmological "constant" and the elapsed proper time (5.41) which is scale-invariant. We can interpret this to mean that "empty" space with a finite curvature can produce a particle with finite mass in accordance with (5.42). This should not be too surprising: empty space with an unseen electromagnetic field can produce particles via pair creation, and there has to be a classical analog of the way in which particles acquire mass via the quantum Higgs mechanism. What is perhaps more surprising is that the vacuum / particle relationship inferred from 5D is compatible with the formal interpretation of wave solutions like (5.43), which can be derived from the straight 4D theory (Section 5.5). It may be possible to realize Mach's Principle within the context of Einstein's general theory of relativity, though this requires some mental flexibility.

The implications of what we have learned in this chapter are widespread. In general, a change in coordinates which involves the extra coordinate $x^4 = l$ will not leave the metric in canonical form, but will instead introduce a significant scalar field via $g_{44} = \pm \, \Phi^2$ where $\Phi = \Phi(x^\gamma, l)$. Then the 5D metric has the form (5.1), with an effective 4D energy-momentum tensor given by (5.2). The latter shows that the source of the gravitational field is a *mixture* of what have hitherto been called matter and vacuum. The conventional split into ordinary material and a vacuum field (measured by a constant Λ) is just the result of suppressing the scalar field. If we wish to measure the energy density of the latter with the traditional symbol, then we should write $\Lambda \, (x^\gamma, \, l \,)$. More cogently, we realize that in 5D the historical division between "matter" and "vacuum" is obsolete.

References

Alvarez, E., Gavela, M.B. 1983, Phys. Rev. Lett $\underline{51}$, 931.

Anderson, J.D., et al. 1998, Phys. Rev. Lett. $\underline{81}$, 2858.

Anderson, J.D., et al. 2002, Phys. Rev. D $\underline{65}$, 082004.

Bertolami, O., Paramos, J. 2004, DF / IST-1. 2003.

Bini, D., Cherubini, C., Mashhoon, B. 2004, Phys. Rev. D $\underline{70}$, 044020.

Birrel, N.D., Davies, P.C.W. 1982, Quantum Fields in Curved Space (Cambridge U. Press, Cambridge).

Csaki, C., Ehrlich, J., Grojean, C. 2001, Nucl. Phys. B 604, 312.

Eddington, A. 1958, The Expanding Universe (U. Michigan Press, Ann Arbor).

Einstein, A. 1922, The Meaning of Relativity (Princeton U. Press, Princeton).

Henry, R.C. 1991, Ann. Rev. Astron. Astrophys. 29, 89.

Hoyle, F. 1949, in Problems of Cosmical Aerodynamics (Int. Union Theor. Appl. Mech. and Int. Astron. Union), 195.

Hoyle, F., Narlikar, J.V. 1974, Action at a Distance in Physics and Cosmology (Freeman, San Francisco).

Huang, W.-H. 1989, Phys. Lett. A 140, 280.

Kalligas, D., Wesson, P.S., Everitt, C.W.F. 1995, Astrophys. J. 439, 548.

Kolb, E.W., Lindley, D., Seckel, D. 1984, Phys. Rev. D 30, 1205.

Liko, T., Wesson, P.S. 2005, Int. J. Mod. Phys. A20, 2037.

Linde, A.D. 1991, Inflation and Quantum Cosmology (Academic Press, Boston).

Liu, H., Wesson, P.S. 1998, Int. J. Mod. Phys. D 7, 737.

Mansouri, F. 2002, Phys. Lett. B 538, 239.

Matute, E.A. 1997, Class. Quant. Grav. 14, 2271.

Mashhoon, B., Liu, H., Wesson, P.S. 1994, Phys. Lett. B 331, 305.

Mashhoon, B., Wesson, P.S. 2004, Class. Quant. Grav. 21, 3611.

Overduin, J.M. 1999, Astrophys. J. 517, L1.

Padmanabhan, T. 1993, Structure Formation in the Universe (Cambridge U. Press, Cambridge).

Padmanabhan, T. 2002, Class. Quant. Grav. 19, L167.

Padmanabhan, T. 2003, Phys. Rep. 380, 235.

Ponce de Leon, J. 2001, Mod. Phys. Lett. A 16, 2291.

Rubakov, V.A., Shaposhnikov, M.E. 1983, Phys. Lett. B 125, 139.

Seahra, S.S., Wesson, P.S. 2001, Gen. Rel. Grav. 33, 1731.

Shiromizu, T., Koyama, K., Torii, T. 2003, Phys. Rev. D 68, 103513.

Strauss, M.A., Willick, J.A. 1995, Phys. Rep. 261, 271.

Tapley, B.D., Bettadpur, S., Ries, J.C., Thompson, P.F., Watkins, M.M. 2004, Science 305 (7), 503.

Weinberg, S. 1980, Rev. Mod. Phys. 52, 515.

Wesson, P.S. 1982, Vistas Astron. 25, 411.

Wesson, P.S. 1991, Astrophys. J. 367, 399.

Wesson, P.S. 1999, Space-Time-Matter (World Scientific, Singapore).

Wesson, P.S., Liu, H. 2001, Int. J. Mod. Phys. D 10, 905.

Wesson, P.S., Seahra, S.S., Liu, H. 2002, Int. J. Mod. Phys. D 11, 1347.

Wesson, P.S. 2005, Astron. Astrophys., 441, 41.

Will, C.M. 1993, Theory and Experiment in Gravitational Physics (Cambridge U. Press, Cambridge).

6. EMBEDDINGS IN $N \geq 5$ DIMENSIONS

"Embed? Don't you mean *in* bed?" (Madonna)

6.1   Introduction

Theories like general relativity can be approached via differential equations or differential geometry. There is, of course, an overlap. But the distinction is traditional, and has to do with whether we wish to use exact solutions of the field equations to study physical problems, or the equations and their associated metrics to study the mathematical properties of manifolds (see the books by Kramer et al. 1980 and Wald 1984). In the present volume we have been mainly concerned with the former approach. However, we now turn to the latter, because the subject of embeddings is of considerable importance if we are to properly understand how 4D general relativity fits into $N \geq 5$D field theory. We gave a primer on Campbell's embedding theorem in Section 1.5, which is sufficient to underpin much of the physics of extended gravity. We now wish to give a deeper account of this and related topics, partly to elucidate the connection between the induced-matter and membrane versions of 5D gravity, and partly to see how these can be extended to even higher dimensions. For those readers who are more interested in physics than mathematics, we note that we will return to practical issues in Chapter 7.

The plan of this chapter is as follows. In Section 6.2 we will quote some relevant results, eschewing proofs which would slow the discussion. Extensive bibliographies on embeddings are available

(Pavsic and Tapia 2000; Seahra and Wesson 2003). Several of the results we will mention were obtained a long time ago, but there is still some disagreement about the strength of embedding theorems and their application to physics. In Section 6.3 we will analyse the general embedding problem, establishing the algebra anew so as not to bias ourselves by previous considerations. These results will be used in Section 6.4 to re-prove the Campbell-Magaard theorem, (our conclusions will agree with the summary in Section 1.5). Then in Sections 6.5 and 6.6 we will apply our general work to the induced-matter and membrane theories, before closing in Section 6.7 with some comments on the implications of embeddings.

6.2   **Embeddings and Physics**

The abstract theory of embeddings dates back almost to the original work of Riemann in 1868. The application of embeddings to physics was implicit in the 5D extensions of general relativity by Kaluza in 1921 and Klein in 1926, and was explicit in Campbell's work which was published also in 1926 (see Chapter 1). The subsequent development of the subject was sporadic, but it is useful to note some of the more significant results.

Thus Tangherlini and others proved in the 1960s that the 4D Schwarzschild solution can only be embedded in a flat manifold of dimensionality $N \geq 6$. But it was not til the 1990s that it was realized that Birkoff's theorem in its conventional form breaks down for $N > 4$. This followed from the discovery of exact solutions of the empty 5D field equations which are 3D spherically-symmetric and static, but have non-Schwarzschild properties. These "solitons" were later extended to the non-static case. The more general problem of how to embed any solution of Einstein's theory in a higher-dimensional flat space showed that it requires $N \geq 10$. This result builds on work by Schlafli, Janet, Cartan and Burstin. The last showed the importance of the Gauss-Codazzi equations as integrability conditions for the embeddings, a subject we will return to below. Campbell, as we have noted, asked about the embedding of 4D spaces of the type used for general relativity in 5D spaces which are Ricci-flat. This is important, because setting the Ricci tensor to zero provides the most basic set of field equations relevant to physics. The theorem that Campbell conjectured was essentially proved much later by Magaard. It is their work which underlies the currently popular approach wherein Einstein's field equations in 4D are viewed as a subset of the Ricci-flat equations in 5D. While it is not particularly strong, the Campbell-Magaard theorem is widely regarded as a kind of algebraic protection for higher-dimensional theories of relativity which reduce to that of Einstein in 4D. It is, however, only a local theorem.

The more difficult problem of global embeddings was considered by Nash, and extended to metrics of indefinite signature by Clarke and Greene. Theories of physics in which spacetime is globally embedded in a flat, higher-dimensional manifold are yet to receive serious attention. But there is no reason why such should not be developed, especially since the theorems have been considered for the cases where the 4D space is compact and non-compact, a difference which some researchers have noted is open to observational test using high-redshift sources like QSOs.

The possibility that higher dimensions might <u>not</u> be compactified was only taken seriously by the physics community in the 1990s. Induced-matter theory dates from 1992, and essentially uses the non-compact extra dimension to give a geometrical account of the origin of matter, as we have discussed at length previously (see Wesson 1992a,b; Wesson and Ponce de Leon 1992). Membrane theory dates from slightly later, and assumes that the non-compact extra dimension is split by a singular hypersurface with Z_2 symmetry, so accounting for why particle interactions in the brane are stronger than gravity which propagates outside it, as we have summarized elsewhere (see Arkani-Hamed, Dimopoulos and Dvali 1998, 1999; Antionadis et al. 1998; Randall and Sundrum 1998, 1999a,b). While these non-compact theories are differently motivated, they are known to possess essentially the same field equations and equations of motion (Ponce de Leon 2001, 2004). Below, we will compare the induced-matter and braneworld models from the perspective of embeddings in 5D.

For dimensionalities higher than 5, some results are known, but only for special values of N. For example, Horava and Witten (1996) showed that the compactification paradigm was not a prerequisite of string theory, which latter is attractive in that it automatically avoids the singularities associated with point particles. They discovered an 11D theory, which has the topological structure of $R^{10} \times S^1 / Z_2$.

This is related to the 10D $E_8 \times E_8$ heterotic string, via dualities. In this theory, the endpoints of open strings reside on a $(3 + 1)$ brane, so standard-model interactions are confined to it, while gravity propagates outside. This is technically a braneworld model, thought it predates the more popular theory with that name outlined above. As another example of $N > 5D$ physics, we can mention $N = 26$ (Bars and Kounnas 1997; Bars, Deliduman and Minic 1999). Models of this type can describe the interaction of a particle and a string, and employ a two-time metric (see Section 3.4 for the 5D two-time metric). The double nature of the timelike dimension can admit supersymmetry and various dualities, and shows that the concept of dimensionality can be taken as far as we wish.

However, even if we let the dimensionality run, the physics we obtain is to a certain degree restrained. This is because we have to take notice of the embedding, to which we now turn our closer attention.

6.3   The Algebra of Embeddings

In this section, we will study the properties of ND Riemann spaces anew, without preconceptions concerning the signature or whether there is a singular surface (as for membrane theory) or not (as for induced-matter theory). For our purposes, the main geometrical object of such a space is the Ricci tensor, which has $(N / 2)(N + 1)$ independent components. We will be largely concerned with spaces whose dimensions differ by 1, and if there is a danger of confusion we will use a hat to denote the higher space, as for example $\hat{R}_{AB}$ versus $R_{\alpha\beta}$. Since our coordinates are numbered from zero and run to n, the total dimensionality of the space is $N = 1 + n$. Then uppercase Latin indices run $0...n$, while lowercase Greek indices run $0...(n-1)$. Square brackets on indices will indicate antisymmetrization. The covariant derivative in the higher space will be denoted by ∇_A, while that in the lower space will be denoted by a semicolon as usual. The Lie derivative will be indicated by £ with an appropriate subscript to identify the dimensionality. We will mean by M any general manifold, in which however we will often be interested in a hypersurface Σ_l defined by the "extra" coordinate l. In the case where Σ_l has some special properties we will denote it by Σ_0, which then in the 4D case is shorthand for spacetime. We are aware that this notation, while standard, is cumbersome. As some small relief, we will adopt geometrical units, so that the speed of light and Newton's gravitational constant are unity.

On our $(n + 1)$-dimensional manifold (M, g_{AB}) we place a coordinate system $x \equiv \{x^A\}$. In our working, we will allow for two possibilities: either there is one timelike and n spacelike directions tangent to M, or there are two timelike and $(n-1)$ spacelike directions tangent to M. The scalar function

$$l = l(x) \tag{6.1}$$

defines our foliation of the higher-dimensional manifold. If there is only one timelike direction tangent to M, we assume that the vector field n^A normal to Σ_l is spacelike. If there are two timelike directions, we take the unit normal to be timelike. In either case, the space tangent to a given Σ_l hypersurface contains one timelike and $(n - 1)$ spacelike directions. That is, each Σ_l hypersurface corresponds to an n-dimensional Lorentzian spacetime. The normal vector to the Σ_l slicing is given by

$$n_A = \varepsilon \, \Phi \partial_A l \quad , \qquad n^A n_A = \varepsilon \quad . \tag{6.2}$$

Here $\varepsilon = \pm 1$. The scalar Φ which normalizes n^A is known as the lapse function. We define the projection tensor as

$$h_{AB} = g_{AB} - \varepsilon n_A n_B \quad . \tag{6.3}$$

This tensor is symmetric ($h_{AB} = h_{BA}$) and orthogonal to n_A.

We place an n-dimensional coordinate system on each of the Σ_l hypersurfaces $y \equiv \{y^\alpha\}$. The n basis vectors

$$e^A_\alpha = \frac{\partial x^A}{\partial y^\alpha} \quad , \qquad n_A e^A_\alpha = 0 \tag{6.4}$$

are by definition tangent to the Σ_l hypersurface and orthogonal to n^A. It is easy to see that e^A_α behaves as a vector under coordinate transformations on $M \left[\phi : x \to \bar{x}(x) \right]$ and a one-form under coordinate transformations on $\Sigma_l \left[\psi : y \to \bar{y}(y) \right]$. We can use these basis vectors to project higher-dimensional objects into Σ_l hypersurfaces. For example, for an arbitrary one-form on M we have

$$T_\alpha = e_\alpha^A T_A \qquad . \qquad (6.5)$$

Here T_α is said to be the projection of T_A onto Σ_l. Clearly T_α behaves as a scalar under ϕ and a one-form under ψ. The induced metric on the Σ_l hypersurfaces is given by

$$h_{\alpha\beta} = e_\alpha^A e_\beta^B g_{AB} = e_\alpha^A e_\beta^B h_{AB} \qquad . \qquad (6.6)$$

Just like g_{AB}, the induced metric has an inverse:

$$h^{\alpha\gamma} h_{\gamma\beta} = \delta_\beta^\alpha \qquad . \qquad (6.7)$$

The induced metric and its inverse can be used to raise and lower the indices of tensors tangent to Σ_l, and change the position of the space-time index of the e_α^A basis vectors. This implies

$$e_A^\alpha e_\beta^A = \delta_\beta^\alpha \qquad . \qquad (6.8)$$

Also note that since h_{AB} is entirely orthogonal to n^A, we can write

$$h_{AB} = h_{\alpha\beta} e_A^\alpha e_B^\beta \qquad . \qquad (6.9)$$

At this juncture, it is convenient to introduce our definition of the extrinsic curvature $K_{\alpha\beta}$ of the Σ_l hypersurfaces:

$$K_{\alpha\beta} = e_\alpha^A e_\beta^B \nabla_A n_B = \tfrac{1}{2} e_\alpha^A e_\beta^B \pounds_n h_{AB} \qquad . \qquad (6.10)$$

Note that the extrinsic curvature is symmetric ($K_{\alpha\beta} = K_{\beta\alpha}$). It may be thought of as the derivative of the induced metric in the normal direction. This n-tensor will appear often in what follows.

Finally, we note that $\{y, l\}$ defines an alternative coordinate system to x on M. The appropriate diffeomorphism is

$$dx^A = e^A_\alpha dy^\alpha + l^A dl \qquad . \qquad (6.11)$$

Here

$$l^A = \left(\frac{\partial x^A}{\partial l} \right) \qquad (6.12)$$

is the vector which is tangent to lines of constant y^α. We can always decompose 5D vectors into the sum of a part tangent to Σ_l and a part normal to Σ_l. For l^A we write

$$l^A = N^\alpha e^A_\alpha + \Phi n^A \qquad . \qquad (6.13)$$

This is consistent with $l^A \partial_A l = 1$, which is required by the definition of l^A, and the definition of n^A. The n-vector N^α is the shift vector, which describes how the y^α coordinate system changes as we move from a given Σ_l hypersurface to another. Using our formulae for dx^A and l^A, we can write the 5D line element as

$$dS^2 = g_{AB} dx^A dx^B$$

$$= h_{\alpha\beta} \left(dy^\alpha + N^\alpha dl \right)\left(dy^\beta + N^\beta dl \right) + \varepsilon \, \Phi^2 dl^2 \qquad . \qquad (6.14)$$

This reduces to $ds^2 = h_{\alpha\beta} \, dy^\alpha dy^\beta$ if $dl = 0$, a case of considerable physical interest.

Let us now focus on how n-dimensional field equations on each of the Σ_l hypersurfaces can be derived, given that the

$(n + 1)$-dimensional field equations are

$$\hat{R}_{AB} = \lambda g_{AB}, \qquad \lambda \equiv \frac{2\Lambda}{1-n} \quad . \tag{6.15}$$

Here Λ is the "bulk" cosmological constant, which may be set to zero if desired. In what follows we will extend the 4-dimensional usage and call manifolds satisfying equation (6.15) Einstein spaces.

Our starting point is the Gauss-Codazzi equations. On each of the Σ_l hypersurfaces these read

$$\hat{R}_{ABCD} e_\alpha^A e_\beta^B e_\gamma^C e_\delta^D = R_{\alpha\beta\gamma\delta} + 2\varepsilon K_{\alpha[\delta} K_{\gamma]\beta}$$

$$\hat{R}_{MABC} n^M e_\alpha^A e_\beta^B e_\gamma^C = 2K_{\alpha[\beta;\gamma]} \quad . \tag{6.16}$$

These need to be combined with the expression for the higher-dimensional Ricci tensor:

$$\hat{R}_{AB} = \left(h^{\mu\nu} e_\mu^M e_\nu^N + \varepsilon n^M n^N \right) \hat{R}_{AMBN} \quad . \tag{6.17}$$

The components of this tensor satisfy equations which may be grouped strategically by considering the following contractions of (6.15):

$$\hat{R}_{AB} e_\alpha^A e_\beta^B = \lambda h_{\alpha\beta}$$

$$\hat{R}_{AB} e_\alpha^A n^B = 0$$

$$\hat{R}_{AB} n^A n^B = \varepsilon\lambda \quad . \tag{6.18}$$

The first member of these gives $(n / 2) (n + 1)$ equations, the second gives n and the last is a scalar relation. The total number of equations is $(1/2) (n + 1) (n + 2)$.

Putting (6.17) into (6.18) and making use of (6.16) yields the following formulae:

$$R_{\alpha\beta} = \lambda h_{\alpha\beta} - \varepsilon \left[E_{\alpha\beta} + K_\alpha{}^\mu \left(K_{\beta\mu} - K h_{\beta\mu} \right) \right]$$

$$0 = \left(K^{\alpha\beta} - h^{\alpha\beta} K \right)_{;\alpha}$$

$$\varepsilon\lambda = E_{\mu\nu} h^{\mu\nu} \quad . \tag{6.19}$$

In writing down these results, we have made the following definitions:

$$K \equiv h^{\alpha\beta} K_{\alpha\beta} \tag{6.20}$$

$$E_{\alpha\beta} \equiv \hat{R}_{MANB} n^M e_\alpha^A n^N e_\beta^B, \quad E_{\alpha\beta} = E_{\beta\alpha} \quad . \tag{6.21}$$

The Einstein tensor $G^{\alpha\beta} \equiv R^{\alpha\beta} - g^{\alpha\beta} R / 2$ on a given Σ_l hypersurface is given by

$$G^{\alpha\beta} = -\varepsilon \left(E^{\alpha\beta} + K^\alpha{}_\mu P^{\mu\beta} - \tfrac{1}{2} h^{\alpha\beta} K^{\mu\nu} P_{\mu\nu} \right) + \lambda \left[1 - \tfrac{1}{2}(n + \varepsilon) \right] h^{\alpha\beta} ,$$

$$\tag{6.22}$$

where we have defined the (conserved) tensor

$$P_{\alpha\beta} \equiv K_{\alpha\beta} - h_{\alpha\beta} K$$

$$P^{\alpha\beta}{}_{;\beta} = 0 \quad . \tag{6.23}$$

This tensor is essentially the one which appears in other studies. For example, it has the same algebraic properties as the momentum conjugate to the induced metric in the ADM formulation of general relativity (Wald 1984; though note that here the direction orthogonal to Σ_l is not necessarily timelike, so $P_{\alpha\beta}$ is not formally a canonical momentum in the Hamiltonian sense). Alternatively, it is the tensor which appears in the vector sector of the induced-matter theory (Wesson and Ponce de Leon 1992), and which contains the non-linear terms for the matter in membrane theory (Ponce de Leon 2001). To complete our analysis we recall that the condition that the 4D Einstein tensor have zero divergence imposes a condition on $E_{\alpha\beta}$ of (6.21). Making use of the second member of (6.19), this is

$$E^{\alpha\beta}{}_{;\beta} = \varepsilon\left(K_{\mu\nu}K^{\mu\nu;\alpha} - K^{\mu\beta}K^{\alpha}_{\mu;\beta}\right) \quad . \tag{6.24}$$

This may be regarded as a condition, satisfied by the geometric quantities on the Σ_l hypersurfaces described by (6.19), when we ask that the physical quantities associated with matter be conserved.

In fact, the preceding analysis shows that (6.19) are in essence the geometric formulation of the field equations for ND relativity.

We will make use of the field equations below, where we will use them to infer general properties of physics, on the assumption that it is the Σ_l hypersurface which we experience. For now, we remark that it is possible to solve the first member of (6.19) for $E_{\alpha\beta}$ and substitute the result into the last member of those relations. The result is

$$(n-1)\lambda = R + \varepsilon\left(K^{\mu\nu}K_{\mu\nu} - K^2\right) \quad . \qquad (6.25)$$

This scalar relation, together with the n relations of the second member of (6.19), provide $(1+n)$ generalizations of the well-known constraints on the Hamiltonian approach to field theories like general relativity. In the $(1 + 3)$ theory, they are the constraints attendant on the initial-value problem and numerical work on the Einstein equations. In the $(1+4)$ theory, they are similar to the relations found for the Randall-Sundrum braneworld scenario (Shiromizu, Maeda and Sasaki 2000). However, we will in what follows not be so much concerned with applying our algebra to constraints, but more with counting the number of independent relations concerned so as to establish results on embeddings.

6.4   The Campbell-Magaard Theorem

With the algebra we have established, it is possible to re-prove Campbell's theorem, independent of previous considerations (Campbell 1926; Magaard 1963; Lidsey et al. 1997; Wesson 1999). Our approach will be heuristic, serving to point towards the physical applications we will take up in the section following.

In (6.19) we have a set of ND field equations which are the geometrical-language analogs of the physically-motivated ones (6.15). The equations (6.19) are defined on each of the Σ_l hypersurfaces, which are labeled by a coordinate (l) that we expect on physical grounds to prove special compared to the others (y). The essential geometrical objects, which can be thought of as spin-2 fields, are

$$h_{\alpha\beta}(y,l), \quad K_{\alpha\beta}(y,l), \quad E_{\alpha\beta}(y,l) \quad . \quad (6.26)$$

Each of these tensors is symmetric, so there are $3(n/2)(n+1)$ independent dynamical quantities governed by (6.19). For book-keeping purposes, we can regard these components as those of a dynamical super-vector $\Psi^{\alpha} = \Psi^{\alpha}(y, l)$. Now the field equations (6.19) contain no derivatives of the tensors (6.26) with respect to l. This means that the components $\Psi^{\alpha}(y, l)$ must satisfy (6.19) for each and every value of l. In alternative language, the field equations on Σ_l are "conserved" as we move from hypersurface to hypersurface. That is, the field equations (6.19) for $(n+1)$ D are in the Hamiltonian sense constraint equations. While this is important from the formal viewpoint, it means that the original, physical equations (6.15) tell us nothing about how Ψ^{α} varies with l. If so desired, equations governing the l-evolution of Ψ^{α} could be derived in a number of equivalent ways. These include isolating l-derivatives in the expansion of the Bianchi identities $\nabla_A G^{AB} = 0$; direct construction of the Lie derivatives of h_{AB} and $K_{AB} = h_A{}^C \nabla_C n_B$ with respect to l^A; and formally re-expressing the gravitational Lagrangian as a Hamiltonian (with l playing the role of time) to obtain the equations of motion. Because the derivation of $\partial_l \Psi^{\alpha}$ is tedious and not really germane to our discussion, we will omit it from our considerations and turn to the more important problem of embedding.

Essentially, our goal is to find a solution of the higher-dimensional field equations (6.15) or (6.19) such that one hypersurface Σ_0 in the Σ_l foliation has "desirable" geometrical properties. For example, we may want to completely specify the induced metric on, and hence the intrinsic geometry of, Σ_0. Without loss of generality, we can assume that the hypersurface of interest is at $l = 0$. Then to successfully embed Σ_0 in M, we need to do two things:

(a) Solve the constraint equations (6.19) on Σ_0 for $\Psi^\alpha(y, 0)$ such that Σ_0 has the desired properties (this involves physics).

(b) Obtain the solution for $\Psi^\alpha(y, l)$ in the bulk (i.e. for $l \neq 0$) using the evolution equations $\partial_l \Psi^\alpha$ (this is mainly mathematics).

Both of these things have to be achieved to prove the Campbell-Magaard theorem, along with some related issues. Thus, we have to show that (a) is possible for arbitrary choices of $h_{\alpha\beta}$ on Σ_0, and we have to show that the bulk solution for Ψ^α obtained in (b) preserves the equations of constraint on $\Sigma_l \neq \Sigma_0$. The latter requirement is necessary because if the constraints are violated, the higher-dimensional field equations will not hold away from Σ_0. This issue has been considered by several authors, who have derived evolution equations for Ψ^α and demonstrated that the constraints are conserved in quite general $(n + 1)$-dimensional manifolds (Anderson and Lidsey 2001; Dahia and Romero 2001a, b). Rather than dwell on this well-understood point, we will concentrate on the n-dimensional field equations for Σ_0; and assume that, given $\Psi^\alpha\left(y,0\right)$, the rest of the $(n + 1)$D geometry can be generated using evolution equations, with the resulting higher-dimensional metric satisfying the appropriate field equations.

With the preceding issues understood, and the weight of the algebra of the previous section established, the Campbell-Magaard theorem becomes close to obvious. The argument simply involves counting.

From above, we recall that our super-vector Ψ^α has a number of independent dynamical components given by $n_d = (3n\,/\,2)\,(n + 1)$. This can be compared to the number of constraint equations on Σ_0, which as we have seen is $n_c = (1/2)\,(n + 1)\,(n + 2)$. For $n \geq 2$, it is obvious that n_d is greater than n_c, which means that our system of equations is under-determined. The number of free parameters is $n_f \equiv n_d - n_c = (n^2 - 1)$. Therefore, we can freely specify the functional dependence of $(n^2 - 1)$ components of $\Psi^\alpha\,(y, 0)$. In other words, since n_f is greater than the number of independent components of $h_{\alpha\beta}$ for $n \geq 2$, we can choose the line element on Σ_0 to correspond to any n-dimensional Lorentzian manifold and still satisfy the constraint equations. This completes the proof of the theorem: *Any n-dimensional manifold can be locally embedded in an $(n + 1)$-dimensional Einstein space.*

This theorem has been used in the literature in several forms, and there has been some discussion of the physical latitude afforded by the just-noted $(n^2 - 1)$ components of algebraic freedom (see

Section 1.5). This number can arguably be cut down, after the constraints are imposed and the induced metric is selected, by a further $(n/2)(n+1)$, leaving $(n+1)(n/2-1)$ components. However, this still leaves a significant degree of arbitrariness in the embedding problem. It implies, for example, that the same solution on Σ_0 can correspond to different structures for M. Due to this and related concerns, the Campbell-Magaard theorem has sometimes been criticized as weak, a view which seems to us to be unjustified. If it were not for the considerable algebraic apparatus we set up in Section 6.3, it would not be clear that such a theorem exists. Indeed, it is due to that apparatus that a simple counting of degrees of freedom suffices to establish the theorem. As counter-comments to the charge that the theorem is weak, we can point out that it is possible to imagine theories of gravity in which it does not hold; and that it has significant implications for so-called lower-dimensional theories of general relativity, where workers ignorant of the theorem have blithely invented field equations which are mathematically tractable but do not respect the constraints handed down by differential geometry. Our view is pragmatic: Einstein's theory is a highly successful theory of gravity based on Riemannian geometry, and since we know that the real world has at least 4 dimensions, we should use the Campbell-Magaard theorem as a "ladder" to higher dimensions.

6.5   Induced-Matter Theory

In this section, and the following one on membrane theory, we will assume that our (3+1) spacetime is a hypersurface in a 5D manifold. We will therefore be able to use the relations of Sections 6.3 and 6.4 with $N = (n + 1) = 5$. We will also employ some results from the literature on dynamics (Ponce de Leon 2001, 2004; Seahra 2002; Seahra and Wesson 2003). Our aim in this and the following section is to understand the nature of 4D matter in a 5D space which is either smooth or divided.

Induced-matter theory is frequently called space-time-matter (STM) theory, because the extra terms in the 5D Ricci tensor act like the matter terms which balance the 4D Einstein tensor in general relativity (Wesson 1999). The field equations in our current approach are (6.19) with $\lambda = 0$. Despite the fact that the Einstein tensor in 5D is zero, it is finite in 4D and is given by (6.22):

$$G^{\alpha\beta} = -\varepsilon\left(E^{\alpha\beta} + K^\alpha{}_\mu P^{\mu\beta} - \tfrac{1}{2}h^{\alpha\beta}K^{\mu\nu}P_{\mu\nu}\right) \quad . \qquad (6.27)$$

Matter enters into STM theory when we consider an observer who is capable of performing experiments that measure the 4-metric $h_{\alpha\beta}$ or Einstein tensor $G_{\alpha\beta}$ in some neighbourhood of their position, yet is ignorant of the dimension transverse to the spacetime. For general situations, such an observer will discover that the universe is curved, and that the local Einstein tensor is given by (6.27). Now, if this person believes in the Einstein equations $G_{\alpha\beta} = 8\pi T_{\alpha\beta}$, he will be forced to conclude that the spacetime around him is filled with some type of matter field. This is somewhat of a departure from the usual point of view, wherein the stress-energy of matter acts as the source for the curvature of the universe. In the STM picture, the shape of the Σ_l hypersurfaces plus the 5-dimensional Ricci-flat geometry fixes the matter distribution. It is for this reason that STM theory is sometimes called induced-matter theory: the matter content of the universe is induced from higher-dimensional geometry.

When applied to STM theory, the Campbell-Magaard theorem says that it is possible to specify the form of $h_{\alpha\beta}$ on one of the embedded spacetimes, denoted by Σ_0. In other words, we can take any known (3+1)-dimensional solution $h_{\alpha\beta}$ of the Einstein equations for matter with stress-energy tensor $T_{\alpha\beta}$ and embed it on a hypersurface in the STM scenario. The stress-energy tensor of the induced matter on that hypersurface Σ_0 will necessarily match that of the (3+1)-dimensional solutions. However, there is no guarantee that the induced matter on any of the other spacetimes will have the same properties.

We now wish to expand the discussion to include the issue of observer trajectories in STM theory. To do this, we will need the covariant decomposition of the equation of motion into parts tangential and orthogonal to spacetime. These are given by Seahra and Wesson (2003, p.1338). We are mainly interested in the motion perpendicular to Σ_l, when the acceleration is:

$$\ddot{l} = \frac{\varepsilon}{\Phi}\left(K_{\alpha\beta}u^\alpha u^\beta + \mathfrak{F}\right) - \dot{l}\left[2u^\beta\left(\ln\Phi\right)_{;\beta} + \dot{l}n^A\nabla_A\Phi\right] \quad . \quad (6.28)$$

Here l is the extra coordinate, Ψ is the lapse introduced above, and $\mathfrak{F} \equiv n_A f^A$ is the magnitude of any non-gravitational forces (per unit mass) f^A which operate in the higher space. The other symbols are as defined in Section 6.3. (An overdot denotes differentiation with respect to proper time or some other affine parameter.) Equation (6.28) is the covariant version of other relations for 5D dynamics, for example (3.18). It clearly shows that the acceleration perpendicular to Σ_l (given by $\ddot{l}$) is coupled to the motion *in* it (given by the 4-velocity u^α). We proceed to consider 3 cases, specified by the extrinsic curvature $K_{\alpha\beta}$ and the non-gravitational force $\mathfrak{F}$:

(a) $K_{\alpha\beta} \neq 0$ *and* $\mathfrak{F} = 0$. A sub-class of this case corresponds to freely-falling observers. We cannot have $l =$ constant as a solution of the l-equation of motion (6.28) in this case, so observers cannot live on a single hypersurface. Therefore, if we construct a Ricci-flat 5D manifold in which a particular solution of general relativity is embedded on Σ_0, and we put an observer on that hypersurface, then he will inevitability move in the l direction. Therefore, the properties of the induced matter that the observer measures may match the predictions of general relativity for a brief period of time, but this will not be true in the long run. Therefore, STM theory predicts observable departures from general relativity.

(b) $K_{\alpha\beta} = 0$ *and* $\mathfrak{F} = 0$. Again, this case includes freely-falling observers. Here, we can solve equation (6.28) with $dl / d\lambda = 0$ (where λ is an affine parameter). That is, if a particular hypersurface

Σ_0 has vanishing extrinsic curvature, then we can have observers with trajectories contained entirely within that spacetime, provided $\mathfrak{F} = 0$. Such hypersurfaces are called geodesically complete because every geodesic on Σ_0 is also a geodesic of *M*. If we put $K_{\alpha\beta} = 0$ into (6.27), we get the Einstein tensor on Σ_l as $G_{\alpha\beta} = 8\pi T_{\alpha\beta} = -\varepsilon E_{\alpha\beta}$. This implies by (6.24) that the trace of the induced energy-momentum tensor must be zero. Assuming that $T_{\alpha\beta}$ can be expressed as that of a perfect fluid, this implies a radiation-like equation of state. Hence, it is impossible to embed an *arbitrary* spacetime in a 5D vacuum such that it is geodesically complete. This is not surprising, since we have already seen that we cannot use the Campbell-Magaard theorem to choose both $h_{\alpha\beta}$ and $K_{\alpha\beta}$ on Σ_0 – we have the freedom to specify one or the other, but not both. If we *do* demand that test observers are gravitationally confined to Σ_0, we place strong restrictions on the geometry and are obliged to accept radiation-like matter.

(c) $K_{\alpha\beta} \neq 0$ *and* $\mathfrak{F} = -K_{\alpha\beta}u^{\alpha}u^{\beta}$. In this case, we can solve (6.28) with $dl / d\lambda = 0$ and hence have observers confined to the Σ_0 spacetime. However, $\mathfrak{F} = -K_{\alpha\beta}u^{\alpha}u^{\beta}$ is merely the higher-dimensional generalization of the centripetal acceleration familiar from elementary mechanics. Since we do not demand $K_{\alpha\beta} = 0$ in this case, we can apply the Campbell-Magaard theorem and have any type of induced matter on Σ_0. However, the price to be paid for this is the inclusion of a non-gravitational "centripetal" confining force, whose origin is obscure.

In summary, we have shown that the Campbell-Magaard theorem guarantees that we can embed any solution of general relativity on the spacetime hypersurface Σ_0 of the 5D manifold used by STM or induced-matter theory. However, for pure gravity, particles only remain on Σ_0 if $K_{\alpha\beta} = 0$, which means that the induced matter has $T^\alpha_\alpha = 0$, which implies a radiation-like equation of state. They could be constrained to Σ_0 if there were non-gravitational forces acting, but in general particles which are not photons will wander away from any given slice of spacetime. This confirms what we found in Chapter 3, where (3.18) has no solution for l = constant, in the absence of a fifth force which would violate the Weak Equivalence Principle. On this basis, the departure of particles from Σ_0 discussed here is equivalent to the change over cosmological times in their masses discussed before.

6.6   Membrane Theory

This exists in several forms, all treatable with our preceding algebra. The simplest form imagines one, thin brane. This divides a bulk 5D manifold into two parts separated by a singular hypersurface, which we call spacetime. Gravity propagates outside the brane, but other physics is concentrated on it, largely by virtue of the assumption that there is a significant, negative cosmological constant (i.e., the bulk is AdS_5). This model is simple, but perhaps limited as regards what can be expected for physics, since the latter is automatically reproduced in almost standard form on the membrane. By contrast, the addition of a second, thin brane leads to effects on the branes which are dependent on the intervening AdS_5 space. These effects have to do with the characteristic energies of gravitational versus other interactions, and can lead to a better understanding of the masses of elementary particles. Other results may be obtained by considering more branes, ones which collide, and thick branes. We will concentrate on the original form of the theory.

The hypersurface Σ_0 is located at $l = 0$, about which there is symmetry (Z_2). Since the membrane is thin, the normal derivative of the metric (the extrinsic curvature) is discontinuous across it. This is like the thin-shell problem in general relativity, and we can take over the same apparatus, including the standard Israel junction conditions. These imply that the induced metric on the Σ_l hypersurfaces must be continuous, so that the jump there is zero:

$$\left[h_{\alpha\beta} \right] = 0 \quad . \tag{6.29}$$

This uses the common notation that $X\pm \equiv \lim\left(l \to 0^\pm\right)X$ and $[X] = X^+ - X^-$. In addition, the Einstein tensor of the bulk is given by

$$\hat{G}_{AB} = \Lambda g_{AB} + \kappa_5^2 T_{AB}^{(\Sigma)}$$

$$T_{AB}^{(\Sigma)} = \delta\left(l\right) S_{\alpha\beta} e_A^\alpha e_B^\beta \quad . \tag{6.30}$$

Here the 4-tensor $S_{\alpha\beta}$ is defined via

$$\left[K_{\alpha\beta}\right] \equiv -\kappa_5^2 \varepsilon \left(S_{\alpha\beta} - \tfrac{1}{3} S h_{\alpha\beta}\right) \quad , \tag{6.31}$$

where κ_5^2 is a 5D coupling constant and $S = h^{\mu\nu} S_{\mu\nu}$. The interpretation is that $S_{\alpha\beta}$ is the stress-energy tensor of the standard fields on the brane. To proceed further, we need to invoke the Z_2 symmetry. This essentially states that the geometry on one side of the brane is the mirror image of the geometry on the other side. In practical terms, it implies

$$K_{\alpha\beta}^+ = -K_{\alpha\beta}^-, \quad \left[K_{\alpha\beta}\right] = 2K_{\alpha\beta}^+ \quad . \tag{6.32}$$

Then we obtain

$$S_{\alpha\beta} = -3\varepsilon\kappa_5^{-2} K_{\alpha\beta}^+ \quad . \tag{6.33}$$

This implies that the stress-energy tensor of conventional matter on the brane is entirely determined by the extrinsic curvature of Σ_0 evaluated in the $l \to 0$ limit.

This is an interesting result, because it shows that even if spacetime is a singular hypersurface, what we call ordinary matter depends on how that hypersurface is embedded in a larger world. However, the result could have been inferred from the STM approach (where there is no membrane and matter is a result of the extrinsic metric) plus the junction conditions (which imply that even if there is a membrane the metric is continuous). This correspondence has been commented on in the literature, and has the happy implication that the extensive inventory of exact solutions for induced-matter theory (Wesson 1999) can be taken over to membrane theory. Indeed, we encountered an example of this in Section 2.4, where we noted that a 5D bouncing cosmology could be interpreted as a (4+1)D membrane model. A detailed account of this kind of problem is given elsewhere (Seahra and Wesson 2003 pp.1334 – 1337), along with comments on multiple and thick branes. For a single, thin brane it should be noted that the derivation of models with Z_2 symmetry is helped by the fact that the constraint equations (6.19) are invariant under $K_{\alpha\beta} \rightarrow -K_{\alpha\beta}$. This can be used to construct an algorithm for the generation of braneworld models.

The field equations on the brane are given by (6.22) with $K_{\alpha\beta}$ evaluated on either side of Σ_0. Usually, equation (6.33) is used to eliminate $K_{\alpha\beta}^{\pm}$, which yields the following expression for the Einstein 4-tensor on Σ_0:

$$G_{\alpha\beta} = \frac{\kappa_5^4}{12}\left[SS_{\alpha\beta} - 3S_{\alpha\mu}S^{\mu}{}_{\beta} + \left(\frac{3S^{\mu\nu}S_{\mu\nu} - S^2}{2} \right)h_{\alpha\beta} \right]$$

$$-\varepsilon E_{\alpha\beta} - \frac{1}{2}\lambda(2+\varepsilon)h_{\alpha\beta} \qquad . \qquad (6.34)$$

Since this expression is based on the equations of constraint (6.19), it is entirely equivalent to the STM expression (6.27) when $\lambda = 0$. However, it is obvious that the two results are written in terms of different quantities. To further complicate things, the braneworld field equations are often written with the stress-energy as the sum of a part proportional to the cosmological constant for the hypersurface and another part,

$$S_{\alpha\beta} = \tau_{\alpha\beta} - \bar{\lambda} h_{\alpha\beta} \quad , \tag{6.35}$$

which however is non-unique. On the other hand, the STM field equations are often written in a non-covariant form, where partial derivatives of the induced metric with respect to l appear explicitly instead of $K_{\alpha\beta}$ and $E_{\alpha\beta}$. We believe that this disconnect in language is responsible for the fact that some workers have yet to realize the substantial overlap between induced-matter theory and membrane theory.

As we did in the previous section, let us now turn our attention to observer trajectories. We will use the same equation for the motion in the direction perpendicular to the hypersurface, namely (6.28), which is common to all 5D embeddings. However, to simplify things we will set the lapse via $\Phi = 1$ (this is a 5D gauge choice, so our 4D results will be independent of it). Then in the l direction, the acceleration reads

$$\ddot{l} = \varepsilon \left(K_{\alpha\beta} u^{\alpha} u^{\beta} + \mathfrak{F} \right) \quad , \tag{6.36}$$

where as before $\mathfrak{F}$ denotes the magnitude of non-gravitational forces. By using (6.31), the last relation gives

$$K^{\pm}_{\alpha\beta} u^{\alpha} u^{\beta} = \mp \tfrac{1}{2} \varepsilon \kappa_5^2 \left[S_{\alpha\beta} u^{\alpha} u^{\beta} - \tfrac{1}{3} \left(\kappa - \varepsilon \dot{l}^2 \right) S \right] \quad . \tag{6.37}$$

We can view this as the zeroth-order term in a Taylor-series expansion of $K_{\alpha\beta} u^{\alpha} u^{\beta}$ in powers of l. In this spirit, the acceleration can be rewritten as

$$\ddot{l} = -\tfrac{1}{2}\operatorname{sgn}(l)\kappa_5^2 \left[S_{\alpha\beta}u^\alpha u^\beta - \tfrac{1}{3}\left(\kappa - \varepsilon \dot{l}^2\right)S \right] + \varepsilon\mathfrak{F} + O(l) \quad , \quad (6.38)$$

$$\text{where} \qquad \operatorname{sgn}(l) = \begin{cases} +1 & l > 0 \\ -1 & l < 0 \\ \text{undefined} & l = 0 \end{cases} \qquad . \quad (6.39)$$

Here we are using $u^A u_A = \kappa$ (we assume that u^A is timelike). From (6.38), it is obvious that freely-falling observers $(\mathfrak{F} = 0)$ can be confined to a small region around the brane if

$$S_{\alpha\beta}u^\alpha u^\beta - \tfrac{1}{3}\left(\kappa - \varepsilon \dot{l}^2\right)S > 0 \quad . \qquad (6.40)$$

Of course, if the quantity on the left is zero or the coefficient of the $O(l)$ term in (6.38) is comparatively large, we need to look at the sign of the latter term to decide if the particle is really confined. To get at the physical content of (6.40), let us make the low-velocity approximation $\dot{l}^2 \ll 1$. With this assumption, (6.40) can be rewritten as

$$\int dy \left\{ T_{AB}^{(\Sigma)} - \tfrac{1}{3} Tr \left[T^{(\Sigma)} \right] g_{AB} \right\} u^A u^B > 0 \quad . \qquad (6.41)$$

This is an integrated version of the 5D strong energy condition as applied to the brane's stress-energy tensor, which includes a vacuum energy contribution from the brane's tension. Its appearance in this context is not particularly surprising, since the Raychaudhuri equation asserts that matter which obeys the strong energy condition will gravitationally attract test particles. What we have shown is that slowly-moving test observers can be gravitationally bound to a small region around Σ_0 if the total matter-energy distribution on the brane obeys the 5D strong energy condition.

Finally, we would like to show that the equation of motion (6.38) has a sensible Newtonian limit. Let us demand that all components of the particle's velocity in the spacelike directions be negligible. Let us also neglect the brane's tension and assume that the density ρ of the confined matter is much larger than any of its principle pressures. Under these circumstances we have

$$S_{\alpha\beta}u^\alpha u^\beta \simeq \rho, \qquad h^{\alpha\beta}S_{\alpha\beta} \simeq \kappa\rho \qquad . \tag{6.42}$$

The 5D coupling constant κ_5^2 is taken to be

$$\kappa_5^2 = \tfrac{3}{2}V_3 G_5 \qquad , \tag{6.43}$$

where V_3 is the dimensionless volume of the unit 3-sphere and G_5 is the 5D "Newton" constant. With these approximations, we get the acceleration for freely-falling observers:

$$\ddot{l} \approx -\tfrac{1}{2}\mathrm{sgn}(l)V_3 G_5 \rho + O(l) \qquad . \tag{6.44}$$

This is precisely the result we would obtain from a Newtonian calculation of the gravitational field close to a 3D surface layer in a 4D space, if we used the Gauss law in the form

$$-\int_{dV} \boldsymbol{g} \cdot d\boldsymbol{A} = V_3 G_5 \int_V \rho \, dV \quad . \qquad (6.45)$$

Here the integration 4-volume is a small "pill-box" traversing the brane. Thus we learn that the full general-relativistic equation of motion in the vicinity of the brane (6.38) reduces to the 4D generalization of a known result from 3D Newtonian gravity in the appropriate limit.

In summary, we have seen that the Campbell-Magaard theorem says that we can embed any solution of 4D general relativity in a 5D space with a membrane. However, the matter associated with the brane is not then freely specifiable. This should not be considered a serious concern, though. Particles near the brane will be confined to a small region near it if the 5D strong energy condition is satisfied, and their motions are Newtonian in the appropriate limit.

6.7   Conclusion

To Newton, it would probably have appeared strange to suggest that ordinary 3D space should be embedded in a 4D spacetime continuum. But Einstein showed that a fourth dimension actually simplifies physics while also extending its scope. The current situation is intriguing: we do not know if all of the consequences of a fifth dimension will prove to be desirable, even though 5D relativity exists in two apparently acceptable versions. Induced matter (or space-time-matter) theory explains 4D matter as a geometrical consequence of the fifth dimension, like when we view a movie projected onto a 2D screen from an unperceived depth. Membrane theory views 4D spacetime and its contents as a special surface in the fifth dimension, like when we walk across the 2D surface of the Earth without knowledge of the underlying geology.

In this chapter, we have looked at the possible connections between embeddings and physics (Section 6.2); developed the algebra which necessarily attaches to embeddings if we are to relate higher dimensions to what we already know (Section 6.3); and argued that the Campbell-Magaard theorem, despite its weaknesses, represents the ladder which allows us to move up or down between dimensions (Section 6.4). The induced-matter and membrane theories are fraternal twins, in the sense that they share the same algebra but have different physical motivations (Sections 6.5, 6.6). Both depend on the fifth dimension, and ascribe real meaning to it.

If the fifth dimension becomes a standard part of physics, there is no reason why we should not employ embedding theory to proceed to even higher levels.

References

Anderson, E., Lidsey, J.E. 2001, Class. Quant. Grav. <u>18</u>, 4831.

Antoniadis, J., Arkani-Hamed, N., Dimopoulos, S., Dvali, G.R. 1998, Phys. Lett. B <u>436</u>, 257.

Arkani-Hamed, N., Dimopoulos, S., Dvali, G.R. 1998, Phys. Lett. B <u>429</u>, 263.

Arkani-Hamed, N., Dimopoulos, S., Dvali, G.R. 1999, Phys. Rev. D 59, 086004.

Bars, I., Kounnas, C. 1997, Phys. Rev. D 56, 3664.

Bars, I., Deliduman, C., Minic, D. 1999, Phys. Rev. D 59, 125004.

Campbell, J. E. 1926, A Course of Differential Geometry (Clarendon, Oxford).

Dahia, F., Romero, C. 2001a, gr-qc / 0109076.

Dahia, F., Romero, C. 2001b, gr-qc / 0111058.

Horava, P., Witten, E. 1996, Nucl. Phys. B 475, 94.

Kramer, D., Stephani, H., Herlt, E., MacCallum, M., Schmutzer, E. 1980, Exact Solutions of Einstein's Field Equations (Cambridge U. Press, Cambridge).

Magaard, L. 1963, Ph.D. Thesis (Kiel).

Lidsey, J.E., Romero, C., Tavakol, R., Rippl, S. 1997, Class. Quant. Grav. 14, 865.

Pavsic, M., Tapia, V. 2000, gr-qc / 0010045.

Ponce de Leon, J. 2001, Mod. Phys. Lett. A 16, 2291.

Ponce de Leon, J. 2004, Gen. Rel. Grav. 36, 1335.

Randall, L., Sundrum, R. 1998, Mod. Phys. Lett. A 13, 2807.

Randall, L., Sundrum, R. 1999a, Phys. Rev. Lett. 83, 3370.

Randall, L., Sundrum, R. 1999b, Phys. Rev. Lett. 83, 4690.

Seahra, S.S. 2002, Phys. Rev. D 65, 124004.

Seahra, S.S., Wesson, P.S. 2003, Class. Quant. Grav. 20, 1321.

Shiromizu, T., Maeda, K., Sasaki, M. 2000, Phys. Rev. D 62, 024012.

Wald, R.M. 1984, General Relativity (U. Chicago Press, Chicago).

Wesson, P.S. 1992a, Astrophys. J. <u>394</u>, 19.

Wesson, P.S. 1992b, Phys. Lett. B <u>276</u>, 299.

Wesson, P.S., Ponce de Leon, J. 1992, J. Math. Phys. <u>33</u>, 3883.

Wesson, P.S. 1999, Space-Time-Matter (World Scientific, Singapore).

7. PERSPECTIVES IN PHYSICS

"Given for one instance an intelligence which could comprehend all the forces by which nature is animated and the respective positions of the beings which compose it, if moreover this intelligence were vast enough to submit these data to analysis, it would embrace in the same formula both the movements of the largest bodies in the universe and those of the lightest atom; to it nothing would be uncertain, and the future as the past would be present to its eyes" (Pierre Laplace)

This quotation, while lacking the pithiness of a contemporary sound-bite, still condenses much of what preoccupies physicists. And whether we agree with Laplace (1812) or not, the issues he raises have resonance with what we have discussed in preceding chapters. Physics, while archival by nature, is not a done deal like the Dead Sea scrolls. It is an evolving subject, and certain questions inevitably recur to occupy the serious student. These include the connection between the macroscopic and microscopic domains, the puzzle of determinacy, and the subjective issue of how much we know (or do not know) at any given stage. It is reasonable to ask in broad terms about the future of physics, given that we have already given a fairly fine-grained account of its present status.

We have looked, by turn, at several issues. In higher than 4 dimensions, field equations should be constructed around the Ricci tensor (Chapter 1). For 5D, this means that the 4D big bang is not merely a singularity but has physics attached to it, even when the higher-dimensional space is flat (Chapter 2). The paths of particles are in general affected by a fifth force in 5D relativity, which however vanishes for the canonical metric, when we recover the Weak Equivalence Principle as a geometric symmetry (Chapter 3). The deterministic laws of 5D when regarded inexactly in 4D result in a Heisenberg-type relation, and imply a mass quantum, with an associated (broken) symmetry between the spin angular momenta and squared masses of gravitationally-dominated systems which has some empirical support from astrophysics (Chapter 4). There is also astrophysical support for a decaying cosmological "constant" of the type allowed by 5D theory, where if we view this parameter as a measure of the 4D energy density of the vacuum we infer that the latter is unstable, though in a manner consistent with Mach's Principle (Chapter 5). The connection between 4D general relativity with matter and 5D theory in apparent vacuum is Campbell's theorem, which is a kind of algebraic ladder that spans manifolds of different dimensionality, and thereby yields the physics of induced-matter and membrane theory (Chapter 6). These two versions of 5D gravity are essentially the same, though motivated respectively by cosmology and particle physics. Both are in agreement with observations. The fact that the two approaches have been viewed differently is largely to do with issues of philosophy and language rather than mathematics.

It was Wittgenstein's opinion that much of modern philosophy is unproductive, because it is concerned with language rather than the meaning which lies behind words. Russell and others pointed out that mathematics is a kind of language, and indeed the natural one for science (see Chapter 1). It has certain advantages. For example, it is sanitary, enabling us to put forth hypotheses about the world without emotion. It is also remarkably exact; and in this regard physicists are fortunate in that they have a common set of algebraic rules – as if science were a grandiose game of chess, played according to inviolate laws on which everyone agrees. (A theoretical physicist who does a faulty calculation is like a concert pianist who plays a bum note: it is discordant, easily detected and usually followed by censure.) But unlike the rigid and therefore somewhat sterile game of chess, physics as based on mathematics has a growing edge, where material can be added to widen its scope.

Imagination, however, is at least as important for advancing physics as a sound knowledge of mathematical technique. Einstein is, of course, frequently quoted in this regard. Unfortunately, modern physics frequently gets bogged down in technicalities: there are too many studies aimed at mere algebra or the addition of a decimal point. Perhaps paradoxically, there are also too many attempts at recreating the universe from time-zero, ignoring established theory and newly-acquired data. It is the middle ground which is the most fertile for advancing physics. This middle ground is also the one which leads the way to new pastures. Current interest in the fifth dimension is a good example of a track which starts in established farmland (Einstein's general relativity) and is already in fresh and fruitful territory.

It was Minkowski who showed how to weld time and space together; but as we have remarked, it was Einstein who realized how inextricably they are intertwined through the Principle of Covariance. The algebra of general relativity can be carried out in *any* system of coordinates, and the standard solutions owe their applicability to the fact that we choose to present them in terms of time and space coordinates that are close to our primitive notions of those quantities. This applies to the Schwarzschild solution for the gravitational field outside the Sun, where the radial coordinate is chosen to agree with the body of knowledge accumulated by that most venerable of sciences, astronomy. It also applies to the Friedmann-Robertson-Walker cosmologies, where the time coordinate is chosen to agree with the one familiar from special relativity, and the body of knowledge which supports this from particle physics. However, relativity really *is* relative: we can if we wish express either of these standard solutions (and others) in different coordinates.

Covariance is both the greatest asset and the greatest drawback of field theory in more than the 4 dimensions of spacetime. It is an asset, insofar as we can change coordinates at will, to assist in finding a solution of the field equations. It is a drawback, insofar as a solution may not correspond to something we recognize from the annals of physics, expressed as it is in terms of our primitive notions of time and space. It can be argued that we are at a rather exceptional stage in the development of the subject. It is not enough to have the mathematical tools necessary to solve *N*D field equations. Also required is the skill to show that they are relevant to the real world.

The question of relevance is central to modern physics, which is preoccupied with finding a unification of the interactions of particle physics and gravity. Let us leave aside, for now, the issue of whether extra dimensions provide the best approach to unification. Let us take it as understood that adding one or more extra dimension does no violence to theories like general relativity (whose field equations do not restrict the dimensionality of the manifold), and that the consequent increased richness of the algebra may be directed at explaining new physical phenomena. Then, a basic issue is whether we accept *all* of the solutions of extended theory as relevant to the real world, or only a subset of them.

This is not a trivial concern. From the practical side, most physicists who earn their living by solving field equations know that solutions frequently turn up which are discarded as being "unphysical". The criteria for this are diverse: a solution may contain an unpalatable geometrical attribute (e.g. a singularity in the manifold), or an unacceptable dynamical result (e.g. a local velocity which exceeds lightspeed), or a strange property of matter (e.g. a negative energy density). The act of discarding such solutions is commonplace, but most physicists would prefer that they were not obliged to exercise this kind of scientific censorship. They would rather deal with a theory whose results are always acceptable, by virtue of it being set up to be <u>complete</u>. Milne commented on the need for a theory to be algebraically complete during the early development of Einstein's theory, and other scientists like Dirac were aware of the problem (see Kragh 1990 and Chapter 1). Unfortunately, general relativity is not complete in the noted sense. The most obvious instance of this concerns the energy-momentum tensor. Some information on this usually has to be imported from the non-gravitational domain in order to obtain a solution. It might, for example, be the equation of state, as determined by atomic physics for a classical fluid or quantum theory for relativistic particles. Most dimensionally-extended versions of general relativity share this fault. In some cases, the fault is quite blatant. Again using the example of the energy-momentum tensor, the components of this for the extra dimensions are frequently just guessed. The resolution of this problem is, of course, obvious: When we set up the theory, the number of field equations and the number of unknowns should exactly balance.

This criterion is satisfied by induced-matter theory. As we have seen, this involves setting the 5D Ricci tensor to zero, solving for the potentials, and interpreting the solution as one where the fields and the matter are both aspects of the geometry. Hence the alternative name, space-time-matter (STM) theory. This is complete in the sense noted above, by construction. In 5D there are 15 independent components of the Ricci tensor and the same number of components of the metric tensor. There is no extraneous energy-momentum tensor as such. Rather, the properties of matter are contained implicitly in the first 10 components of the field equations, which by Campbell's theorem can be written as the Einstein equations with sources. The other 5 components of the field equations can be written as 4 conservation laws plus a scalar wave equation. The 15 potentials, following the traditional view, are related to gravitation (10), electro-magnetism (4) and a scalar field (1). This theory unifies the physics normally associated with Einstein and Maxwell, and it is logical to infer that its scalar field is related to the Higgs field of quantum mechanics and determines the masses of particles. It is a unified theory of matter in 5D.

There is nothing sacrosanct, however, about 5D. Indeed, if we are to incorporate the weak and strong interactions of particles – along with their internal symmetry groups – we should consider $N > 5$ dimensions.

In such a theory, we should again proceed by setting the Ricci tensor to zero in order to obtain field equations. (In the absence of an energy-momentum tensor, there is no ambiguity about whether the Ricci tensor or the Einstein tensor should be set to zero, since the one thing implies the other.) Then we can again use Campbell's theorem to go between dimensions, and thereby identify the lower-dimensional sources that are induced by the higher-dimensional vacuum. While this is easy to state, however, it may not be so easy to accomplish. We remarked above that covariance is both the strength and the weakness of ND field theory. We also remarked that space and time are primitive sense concepts, which for that reason we use as the independent variables in our theories. (The dependent quantities are usually more abstract in nature, such as potentials.) But to solve the field equations in theories of this type, it is necessary to make a starting assumption about the form of the metric (or distance measure). That is, it is necessary to make a choice of coordinates, or gauge. In 4D, we are helped in this by our knowledge of how time and space "behave" in certain circumstances. In 5D, we are helped by our knowledge of mechanics, which leads to the canonical metric and the inference that the fifth dimension is related to mass. (See Chapters 1 and 3. For membrane theory as opposed to STM, we are helped by our knowledge of the hierarchy problem to the warp metric, with the inference that the extra coordinate measures distance from the brane.) In a theory with more than 5 dimensions, however, we are in uncharted territory. How can we make a sensible choice of gauge when we do not even have a clear idea of the nature of the higher coordinates?

This is a question which haunts workers in ND field theory. It is largely our ignorance of the nature of the higher coordinates which is responsible for the plethora of papers in the subject, most of which are long on algebra but short on physics. However, it should be recalled that the Covariance Principle is essentially a statement about the arbitrariness of coordinates. As such, it applies even to our standard labels of space and time. At the risk of making a vice into a virtue, it is instructive to reconsider these referents.

Time is probably more discussed than any other concept in physics. This is because everybody has a sense of its passing in the macroscopic world, and because it has to be handled carefully to make sense of the equations which describe the microscopic world. The nature of time has been treated extensively in the books and review articles by Gold (1967), Davies (1974, 2005), Whitrow (1980), McCrea (1986), Hawking (1988), Landsberg (1989), Zeh (1992) and Wesson (1999, 2001). These sources take as their starting point the view of Newton, who in *Principia* (Scholium I) stated that "Absolute, true and mathematical time, of itself, and from its own nature, flows equably without relation to anything external, and by another name is called duration." This sentence is often quoted in the literature, and is widely regarded as being in opposition to the nature of time as embodied later in relativity. However, prior to that sentence, Newton also wrote about time and space that "...the common people conceive these quantities under no other notions but from the relation they bear to sensible objects." Thus Newton was aware that the "common" people in the 1700s held a view of time and other physical concepts which was essentially the same as the one used by Einstein, Minkowski, Poincaré and others in the 1900s as the basis for relativity.

Nowadays, a common view is that the macroscopic arrow of time is set by the evolution of the universe (Gold 1967; Davies 1974; Whitrow 1980). This sounds plausible, given that the universe started in a big bang and has a thermodynamic direction thereby stamped on it. However, closer inspection shows that this view is flawed, because in the comoving frame of standard cosmology we can set the relative velocities of the galaxies to zero (see Chapters 1 and 3). This implies that the cosmological arrow of time is gauge-dependent. It is also difficult to see how events at the distances of QSOs – even if they are temporally directed – can have an influence locally which explains our human sense of the passage of time. For this, we must look to biological processes. In this regard, there appears to be a dilemma: biological processes, such as the building of the human genome, tend to create information over time; whereas the associated thermodynamical processes still produce entropy (Davies 2005). Insofar as information is negative entropy, there is the suspicion of a paradox. The same suspicion, it should be noted, hangs over other physical processes which tend to create order out of chaos, such as the operation of gravity to produce a structured Earth from an amorphous protostellar cloud.

Due to ambiguities of this type, other views on the nature of time have become popular in recent years, wherein the temporal sense is not simply linked to the increase of entropy. One of these is the many-worlds interpretation of quantum mechanics due originally to Everett (1957). In this, all outcomes of quantum-mechanical processes are possible, but we are only aware of those which we call reality. This view, according to several workers including De Witt (1970), is both mathematically and physically consistent. This approach does not directly account for the human perception of the passage of time, but Penrose (1989) has suggested that quantum effects might be amplified by the brain to the level at which they become noticeable.

Another view is that time is an illusion, in the sense that it is an *ordering* mechanism imposed by the human brain on a world in which events actually happen simultaneously. This idea may sound precocious, and leads to the question: Is time instantaneous? However, it can be formulated in a meaningful way (Wesson 2001), and has been taken up independently by a number of deep thinkers. Thus from Einstein as reported by Hoffman (1972): "For us ... the distinction between past, present and future is only an illusion, albeit a stubborn one." While a parallel opinion is from Ballard (1984): "Think of the universe as a simultaneous structure. Everything that's ever happened, all the events that *will* ever happen, are taking place together." And from Hoyle (1963, 1966): "All moments of time exist together," and "If you were aware of your whole life at once it would be like playing a sonata simply by pushing down all the notes on the keyboard." The latter author points out that this view of time need not be mystical. He considers a 4D world with coordinates (t, xyz) and a surface defined by $\phi(t, xyz) = C$, where "We could be said to live our lives through changes of C." This approach to time clearly has an overlap with the many-worlds interpretation of Everitt noted above. It is basically saying that reality is simultaneous, and that time is our way of separating events in it.

The concept of simultaneity is gauge invariant, in the sense that a null interval remains so no matter how we change the coordinates, including the time. In 4D special relativity, $ds = 0$ specifies zero separation, which we interpret to mean that particles in ordinary

3D space exchange photons along straight paths. (Here ds as the element of "proper" time takes into account the velocity in ordinary 3D space, as embodied in the Lorentz transformations, without which particle accelerators would be mere junk.) In 4D general relativity, we use the same prescription to argue that in the presence of gravitational fields, particles are connected by light rays which follow curved paths. In 5D relativity, as developed in earlier chapters of this book, we used the null path $dS = 0$ to describe the paths of all particles, whether massive or not. The crucial point is that the interval is null, as realized by Einstein in 1905 when he used this as a definition of simultaneity. The rest of the physics flows <u>from</u> this definition.

The notion of a null path is mathematically precise, and is central to modern cosmology. We use it regularly to calculate the size of the horizon, which is the imaginary boundary that separates those particles which are in causal contact from those which are not (Rindler 1977; Wesson 1999). However, the physics which we build around the null path is to a certain extent subjective. We find it difficult to think in a 4D manner, and prefer to split spacetime into its (3+1) component parts. This is why we do not normally say that $ds = 0$ means that particles have zero separation in 4D, but instead say that a photon propagates in time across a portion of ordinary 3D space. It is even harder to think in a 5D manner, so we again find it convenient to decompose the manifold into its component parts, and imagine some influence which propagates through them. However, while the notion of a null path is independent of how we choose coordinates, the splitting of the manifold is not. To this extent, the physics contains a subjective element.

Eddington was the first person of stature to suggest that physics might, at least in part, be subjective. He wrote extensively on this at a time when it was an unpopular view. His book *The Philosophy of Physical Science*, which came out in 1939, was criticized by both physicists and philosophers. Yet a person who reads Eddington with an open mind cannot but find his arguments compelling; and recent developments in cosmology make his views more palatable now than they were before. (There has been a comparable increase in the acceptability of counter-intuitive consequences of quantum mechanics, as considered for example by Bell 2004.) It would be inappropriate to go here into the details of Eddington's philosophy, especially as recent reviews are available (Batten 1994; Wesson 2000; Price 2004). But he basically held the view that while an external world exists, our perception of it is strongly influenced by the physiological and psychological attributes that make us human. He used the analogy of a fisherman, who notices that all the fish he catches are larger than a certain size, and assumes that this is a fact of nature, whereas it is due to the size of the mesh in the net he is using. As a more physical example, in the situation we considered above – where events are simultaneous in 4D or else connected by photons which travel in time through 3D – Eddington would have taken the view that we had in some sense invented the idea of a photon, in order to give an explanation of the situation in a way which fits our perceptions. He was led to the conclusion that biology is actually the most "valid" form of science, in that there is less obstruction between the data and our *interpretation* of the data. Contrarily, Eddington regarded much of physics as *invented* rather than discovered.

The views of Eddington overlap somewhat with those of Einstein, Ballard and Hoyle on time, which we outlined above. These and others have clearly regarded with some doubt the way in which physics has traditionally been presented. In particular, there has apparently always been some scepticism about the status of coordinates. It is therefore not surprising that modern physics – which is focussed on unification through extra dimensions – finds itself beset with questions of interpretation. For our present concerns, there is one question which is paramount. In loose language, it is just this: Are extra dimensions "real"?

We believe that the answer to this is : Yes, provided they are useful.

Lest this be considered specious, it should be recalled that the history of physics is replete with examples of ideas which were adopted because they were useful, or discarded because they were not. It is better to treat the dynamics of a particle with a 4-vector than to deal separately with its energy and momentum. And the aether had become unworkably complicated before the Michelson-Morley experiment terminated its tenure. In the case of extra dimensions, the requirement is simply that they earn their keep.

We have in this volume concentrated on the case of 5 dimensions, which has been known since the 1920s to provide a means of unifying gravity with electromagnetism. This, apparently, was not enough to earn the fifth coordinate the legitimacy of being "real". However, in the last decade there has been an explosion of new interest in 5D relativity, due largely to how the extra dimension may be used to consider problems related to mass. STM theory views the extra dimension as being all around us in the form of matter, and in the special canonical gauge the extra coordinate is essentially the rest mass of a particle. Membrane theory views the extra dimension as perpendicular to a hypersurface which is the focus of particle interactions, gravity spreading away from this in a manner which helps explain why typical masses are smaller than the Planck one which would otherwise rule. The above-posed question about the "reality" (or otherwise) of the fifth dimension is seen to be connected to how useful it proves to be in describing mass.

In this context, there are some specific questions that need to be addressed:

(a) Is the inertial rest mass of a local particle governed by the scalar potential of a global field? This would be a concrete realization of Mach's Principle. But while we have given special results which show the plausibility of this view, a general demonstration is needed.

(b) If the foregoing conjecture is true, how do we account for the disparate masses of elementary particles which have otherwise identical physical properties? For example, how do we calculate the muon / electron mass ratio?

(c) Given that 5D relativity unifies gravity and electromagnetism, we can use the 5 degrees of coordinate freedom to "turn off" the electromagnetic interaction; so where does the physics of the latter "go"? Insofar as we cannot suppress the 10 gravitational potentials in the same manner, it is reasonable to infer that the scalar field takes up the slack. But if so, a simple system like the hydrogen atom ought to have two equivalent descriptions: one where the structure is due to electromagnetism (like the original Bohr model), and one where the electron orbits are due to the form of the scalar field. We lack the second picture.

(d) The laws of mechanics are different in 5D versus 4D, and if the fifth dimension is related to rest mass, we can ask: Is the extra dimension likely to be revealed by more exact tests of the Weak Equivalence Principle? This is a geometric symmetry of 5D relativity for the induced-matter scenario. But as in particle physics, we expect the symmetry to be broken, in this case at a level dependent on the ratio of the test mass to the source mass. This can be examined by upcoming experiments, such as the satellite test of the equivalence principle, which will use test masses in orbit around the Earth.

(e) By simple arithmetic, there are 5 laws of conservation in a 5D world, but if we take a 4D view then discrepancies are inevitable; so can we detect these? On a microscopic scale, these discrepancies mimic the phenomena usually attributed to Heisenberg's uncertainty principle. But on a macroscopic scale, they should be detectable by studying the dynamics of objects such as galaxies. Cosmology is in an era of precision measurements, and we should be able to assess the dimensionality of the universe using astrophysical data.

(f) The universe may well be flat in 5D (even though it is curved in 4D), so it is natural to ask: can we show that its total energy is zero? This question is easy to formulate, but is difficult to answer in a practical sense, given the non-locality of gravitational binding energy. However, a corollary is that the paths of particles in a 5D universe should be null geodesics, which implies that all of its parts are in contact. This leads us to expect that SETI may be a done deal (though we do not know the mechanism); and that all objects in the universe should have identical properties (irrespective of the nature of the standard horizon). Preliminary studies of the spectroscopic properties of QSOs support this view. In a 5D universe where intervals are null, all mass-related properties of objects should be identically the same, everywhere and for all times.

The preceding half-dozen questions provide practical ways to test for the existence of a fifth dimension. It should be recalled, in this connection, that we do not need to "see" an extra dimension in order to admit that it exists. We do not <u>see</u> time, but few of us would deny that it exists. In closing, we point out that the existence or not of a fifth dimension is pivotal to physics. Because if there is a fifth, there should be more ...

References

Ballard, J.G. 1984, Myths of the Near Future (Triad-Panther, London), 34, 39, 99, 109, 112.

Batten, A. 1994, Quart. J. Roy. Astr. Soc. 35, 249.

Bell, J.S. 2004, Speakable and Unspeakable in Quantum Mechanics (2nd. Ed., Cambridge U. Press, Cambridge).

Davies, P.C.W. 1974, The Physics of Time Asymmetry (U. California Press, Berkeley).

Davies, P.C.W. 2005, Astron. Geophys. 46 (2), 26.

De Witt, B.S. 1970, Phys. Today 23 (9), 30.

Eddington, A.E. 1939, The Philosophy of Physical Science (Cambridge U. Press, Cambridge).

Everett, H. 1957, Rev. Mod. Phys. 29, 454.

Gold, T. (ed.) 1967, The Nature of Time (Cornell U. Press, Ithaca).

Hawking, S.W. 1988, A Brief History of Time (Bantam Press, New York).

Hoffman, B. 1972, Albert Einstein, Creator and Rebel (New American Library, New York), 257.

Hoyle, F., Hoyle, G. 1963, Fifth Planet (Heinemann, London), 5 – 8.

Hoyle, F. 1966, October the First is Too Late (Fawcett-Crest, Greenwich, Conn.), 45-46, 64-69, 150.

Kragh, H., 1990. Dirac: A Scientific Biography (Cambridge U. Press, Cambridge).

Landsberg, P.T., 1989, *in* Physics in the Making (eds. Sarlemijn, A., Sparnaay, M.J., Elsevier, Amsterdam), 131

Laplace, P.S. 1812, Analytical Theory of Probability (Courcier, Paris).

McCrea, W.H. 1986, Quart. J. Roy. Astr. Soc. 27, 137.

Penrose, R. 1989, The Emperor's New Mind (Oxford U. Press).

Price, K. (ed.) 2004, Arthur Stanley Eddington: Interdisciplinary Perspectives (Centre for Research in the Arts, Humanities and Social Sciences, Cambridge U., 10 – 11 March).

Rindler, W. 1977, Essential Relativity (2nd. ed., Springer, Berlin).

Wesson, P.S. 1999, Space-Time-Matter (World Scientific, Singapore).

Wesson, P.S. 2000, Observatory 120 (1154), 59.

Wesson, P.S. 2001, Observatory 121 (1161), 82.

Whitrow, G.J. 1980, The Natural Philosophy of Time (Oxford U. Press, London).

Zeh, H. D. 1992, The Physical Basis of the Direction of Time (Springer, Berlin).

INDEX